THE KILLER'S SHADOW

ALSO BY JOHN DOUGLAS AND MARK OLSHAKER

Mindhunter

Journey into Darkness

Unabomber

Obsession

The Anatomy of Motive

The Cases That Haunt Us

Broken Wings

Law & Disorder

The Killer Across the Table

THE KILLER'S SHADOW

THE FBI'S HUNT FOR A WHITE SUPREMACIST SERIAL KILLER

CASES OF THE FBI'S ORIGINAL MINDHUNTER, BOOK 1

JOHN DOUGLAS AND
MARK OLSHAKER

DEY ST.
An Imprint of WILLIAM MORROW

HarperCollins books may be purchased for educational, business, or sales promotional use. For information, please email the Special Markets Department at SPsales@harpercollins.com.

FIRST EDITION

Designed by Angela Boutin

Abstract background image on title page and part titles © aerial333 //stock.adobe.com

Library of Congress Cataloging-in-Publication Data has been applied for.

ISBN 978-0-06-297976-6 (trade paperback edition)
ISBN 978-0-06-307444-6 (library hardcover edition)

20 21 22 23 24 LSC 10 9 8 7 6 5 4 3 2 1

IN REMEMBRANCE OF

Rebecca Bergstrom
Marion Vera Hastings Bresette
Johnny Brookshire
Dante Evans Brown
Victoria Ann "Vicki" Durian
Theodore Tracy "Ted" Fields
Gerald Gordon
Darrell Lane
Alphonce Manning Jr.
David Lemar Martin III
Mercedes Lyn "Marcy" Masters
Harold McIver
Kathleen Mikula
Johnnie Noyes
Lawrence E. Reese
Nancy Santomero
Toni Lynn Schwenn
Arthur Smothers
William Bryant Tatum
Jesse E. Taylor
Raymond Taylor
Leo Thomas Watkins

May their memory be a blessing and a triumph of love over hate.

"Portrait of a bush-league führer named Peter Vollmer, a sparse little man who feeds off his self-delusions and finds himself perpetually hungry for want of greatness in his diet. And like some goose-stepping predecessors he searches for something to explain his hunger, and to rationalize why a world passes him by without saluting. That something he looks for and finds is in a sewer. In his own twisted and distorted lexicon, he calls it faith, strength, truth. But in just a moment Peter Vollmer will ply his trade on another kind of corner, a strange intersection in a shadowland called... the Twilight Zone."

—Rod Serling, opening monologue from "He's Alive,"
The Twilight Zone, aired January 24, 1963

THE KILLER'S SHADOW

PROLOGUE

The sniper had been meticulous. He'd examined potential targets around the city and then surveilled the landscape around the one he'd finally selected for the optimal position the day before.

The Brith Sholom Kneseth Israel Congregation was located on Linden Avenue in the Richmond Heights suburb of St. Louis. It was close to Interstates 64 and 170, so the getaway would be quick and efficient. Across the street there was a knoll with bushes, high grass, and a telephone pole, which would provide good cover and a clean angle on the synagogue parking lot. It was about a hundred yards away—no problem for his bolt-action, center-fire Remington 700 .30-06 semiautomatic hunting rifle, fitted with a telescopic sight. He had brought it to his chosen position ahead of time, hidden in a black guitar case under the bushes. He'd already taken the precaution of filing down the weapon's serial number so it couldn't be traced—he tried to never use the same weapon twice, just part of his routine planning. He arrived by bicycle so no vehicle could be identified, no tire tracks traced to a particular type of car. He left his own vehicle at a shopping center parking lot, some distance away.

It was October 8, 1977, a mild, sunny Saturday, with autumn just starting to make its arrival felt.

He had hammered two ten-inch nails into the telephone pole when he visited the site on Friday and stretched a sock between them to serve as a gun rest.

Then he waited.

He had looked up the service time and knew that it let out around one o'clock, in time for people to have lunch.

It was just a few minutes after that when the doors opened and the congregants started pouring out. Two men stopped to talk to each other in the parking lot on the north side of the building. There was a young girl standing next to one of them, and a woman and two other girls nearby—wife and children, probably. The first man started getting into his car.

The sniper tensed his grip on the rifle, focused on his heart rate, and controlled his breathing into a conscious, consistent cadence. He peered into the sight and smoothly squeezed off two quick shots in the direction of the two men. There was a loud report that must have sounded like firecrackers exploding to everyone coming out of the synagogue, but he felt, more than heard, it as the concussive wave of the firing lifted the barrel and pushed his shoulder back. A split second later he saw one of the men—the one with the girl by his side—clutch his chest and go down. The other man seemed to flinch, but the sniper couldn't tell whether he'd gotten him or not. People nearby instinctively crouched down or dove to the ground. The second guy quickly snatched up the other man's little girl, who was yelling in terror, and dashed for cover between the parked cars. All hell broke loose as the woman with the two girls rushed to the fallen man and bent over him on the pavement. When she stood up, she was screaming and there was blood all over the front of her dress.

After the shooting started, according to multiple later reports, several children ran back into the synagogue building, where the majority of congregants remained, shouting, "They're shooting people! They're killing people!"

Taking advantage of the pandemonium, the sniper repositioned the rifle on his shoulder, refocused his aim, and fired off three more shots in the general direction of the synagogue building. The bang only increased the panic. He might have hit one more man; he wasn't sure. But now it was time to get the hell out of there.

He quickly but carefully wiped down the rifle and guitar case of any possible fingerprints, placed the rifle inside the case, and threw it into the bushes. Then he climbed onto the bicycle and sped off to the nearby shopping center parking lot, unlocked and climbed into his car, turned the key in the ignition, gunned the accelerator, and took off.

THE MAN HIT WAS NAMED GERALD "GERRY" GORDON, FORTY-TWO YEARS OF age, who, with his wife, Sheila, and three daughters, was among the two hundred or so guests attending the bar mitzvah of Ricky Kalina, son of Maxine and Merwyn Kalina, two of Gerry's closest friends. He had just been congratulating and saying goodbye to Ricky right outside the synagogue doors before heading to his car with Sheila and the girls. Gerry had a reputation as something of a jokester, so when he went down clutching his chest after the loud pops, people around him thought he was playacting. Even Steven Goldman, the friend he was talking to in the parking lot, thought Gerry was joking, until he glanced down and saw the blood spreading across Gerry's chest. That was when he snatched up Gerry's little girl to protect her from whatever was happening.

Ricky ran back into the building to find his parents and get to a phone to call for help. The ambulance arrived within minutes and took Gerry to St. Louis County Hospital, where he was rushed into surgery. The single bullet had pierced his left arm and lodged in his chest. He died on the operating table at around three from blood loss and damage to his lung, stomach, and other organs. He was a salesman for the Ropak Corporation, a paper-products distributing firm. His three daughters were named Hope, Michele, and Traci.

The police also arrived quickly to the scene and immediately cordoned off the area and began interviewing witnesses and looking for physical evidence.

Just as Gordon went down, Steve Goldman thought he felt something like a bug bite on his left shoulder but forgot all about it in the chaos after his friend had been shot. Minutes later, when he was giving his shattering account to a police officer, the officer spotted a bullet hole in Goldman's suit jacket. That was how close, Steve suddenly realized, he had come to sharing Gerry Gordon's fate.

Another man, thirty-year-old William Lee Ash, was struck by one of the final three shots, and he knew he had been hit. He was also treated at County. He lost the little finger on his left hand when the shot first hit the hand and then embedded in his hip. His wife, Susan, was Maxine Kalina's cousin.

The police agreed that five shots were fired in total, in what investigators stated was "a highly premeditated attack." They found the Remington rifle inside the black guitar case not far from the telephone pole into which the two nails had been hammered. The sock was still attached and damp from the previous evening's rain, indicating the shooter had been there before. An empty five-round clip was found in the rifle, and five spent shells were recovered nearby. No fingerprints were detected.

A St. Louis County police helicopter pilot spotted a man running across a highway overpass near Brith Sholom and believed he might have been the sniper. Police recovered one bullet in the radiator of a car parked in the synagogue lot and another lodged in the wall of a home across Linden Avenue from the house of worship. Police speculated that a bicycle found about a block away could have been used by the assailant.

Witnesses noted that a man carrying a black guitar case had been in the vicinity about an hour before the shooting. They described him as about five feet, ten inches tall, of medium build, with long curly hair, a fair complexion, and acne scars. He was estimated to be between nineteen and twenty-five years of age. Other witnesses saw a man running through a shopping center parking lot near the synagogue after the shooting, looking over his shoulder as if afraid he was being pursued.

The attack was a huge story throughout the region. Though area Jewish leaders were quoted in Monday's edition of the *St. Louis Post-Dispatch* as discounting anti-Semitism as a motive in the shooting, most synagogues added guards and security patrols. "The whole city seemed to go into a panic after the sniper incident," a police dispatcher was quoted as saying in the same article.

St. Louis police lieutenant Thomas H. Boulch of the Major Case Squad was assigned to lead the investigation, heading up a team of more than twenty detectives, working with officers from the Richmond Heights PD. Boulch was quoted in the newspaper as saying they were keeping an open mind as to motive. "We don't want to close any avenue. We're following up on kooks, radical groups, anti-Semitic groups, anyone." The squad looked into the possibility that the sniper was after a specific individual.

Brith Sholom rabbi Benson Skeff could think of no reason why his synagogue would be targeted. Norman A. Stack, executive director of the St. Louis Jewish Community Relations Council, agreed, telling the *Post-Dispatch,* "Nothing has really happened recently to indicate any problems within the Jewish community of St. Louis. It could have happened to anyone anywhere."

That same week, the St. Louis County Police Department's Identification Bureau successfully employed a chemical treatment to raise the murder weapon's serial number. With assistance from the Bureau of Alcohol, Tobacco and Firearms in Washington, D.C., the rifle was traced to a previous owner in Irving, Texas. He said he had sold it about four weeks previously for two hundred dollars and provided a description of the purchaser to a police sketch artist. Witnesses said that sketch looked like the man they had seen running through the nearby shopping center parking lot, though investigators were unwilling to dismiss the possibility that two individuals were involved in the sniper attack.

Whoever it was, though—one person or more—had managed to disappear.

PART 1

ON THE HUNT FOR A KILLER

CHAPTER 1

OCTOBER 10, 1980

I was in my office at Quantico when I got the call. Back then, the FBI's Behavioral Science Unit (BSU) was located in a large and dingy basement space under the library, with our individual offices defined by six-foot-high partitions; whenever any of us was on the phone, everyone else in the office would hear the conversation, whether they wanted to or not. There were usually eight of us working down there together, instructors in criminal psychology, sociology, police stress, and other related subjects. At that point, I was the only full-time operational profiler, though I still put in some time on the Criminal Psych classes (officially known as "Criminal

Psychology"), which we were trying to make more directly useful for law enforcement.

I'd come up with the idea of interviewing incarcerated repeat killers and violent predators when my instructor partner Special Agent Robert "Bob" Ressler and I had been out teaching "road schools" with local police departments and law enforcement agencies a few years earlier. These schools were kind of a highlights version of the curriculum we taught at Quantico to new agents and senior officers, and detectives chosen to be FBI Academy fellows.

The initial aim of these prison interviews had been to learn and understand what the offenders said was going on in their minds before, during, and after the crime—along with the behavior they displayed while they were committing the crime—so we could correlate it with the evidence they'd left at the crime scenes and body dump sites. And although the interviews of men like Santa Cruz, California, "Coed Killer" Edmund Kemper, New York City's "Son of Sam" David Berkowitz, Oregon shoe fetish killer Jerome Brudos, and Los Angeles "Helter Skelter" murder master Charles Manson became the foundation for the FBI's behavioral profiling program, the original motivation was just to be better instructors and not look foolish in front of those seasoned detectives and chiefs in the FBI's National Academy program, who often had more experience and direct knowledge of the crimes than we did.

Applying the knowledge gained from our interviews, we were trying to transform Criminal Psych into an actual investigative tool. It was becoming a popular course, and police and sheriff's officers began bringing cases from their hometowns for us to review. In 1979, we received just under sixty cases, and

I started tapering off from full-time teaching. The following year, the caseload more than doubled, which is why, in January 1980, I gave up my full-time teaching assignment and became program manager of the criminal profiling program and a "guest lecturer" at the academy.

The call that morning was from someone I knew well: Dave Kohl, a unit chief working under Joseph P. Schulte Jr., chief of the Civil Rights Division up at FBI headquarters in Washington, D.C. I had known Dave since my days as a street agent in Milwaukee, when I had been on the field office SWAT team and he was the team leader and supervisor of the reactive squad—the squad that handled kidnapping, extortion, and related crimes. A former marine officer and a champion wrestler in college, Dave had been my "rabbi" in the Bureau, an expression we used to describe an older or more senior official who took you under his wing, became your mentor, and generally looked out for you. But it was quickly clear that this was not a personal call.

"John, have you heard of Joseph Paul Franklin?"

"Umm, yeah," I said, quickly trying to associate the name with a case. "Isn't he the guy they think shot those two Black joggers in Salt Lake City?"

"That's the one. And he could also be good for murders of Blacks in Ohio, Indiana, Pennsylvania, maybe some other places."

"Didn't they arrest him and he escaped from a police station or something?"

"Right. In Florence, Kentucky; it's not far from Cincinnati. He just climbed out the window during a break in the interrogation. That's why I'm calling. You'll probably be getting a formal request from headquarters, but I'm asking you now. Do you think you could do some kind of psychological assessment that could help us find him?"

"Well," I said, "I think from research and experience I have a handle on how the criminal mind works, although this is pretty unusual. Normally, we deal with *unknown* subjects and try to describe the UNSUB's characteristics from crime scene evidence, police reports, and victimology. Here, you know *who* the guy is, just not where you can find him or what he'll do next. The elements of an assessment are still the same, though. No guarantees, but I'll give it my best shot. Are you going to be sending materials down here?"

"There's so much, you'd better come up here and pick and choose what you want. How long do you think this will take?"

"How about a couple of days?" I suggested.

"Can we have it in the next twenty-four hours?" Dave countered.

A beat of silence, then, "Okay," I said. "I'll be up this afternoon."

"I'll see you then. And John, there's something else you need to know. This may have been the guy who shot Vernon Jordan, and we know he wrote a threatening letter to President Carter when he was a candidate because of Carter's pro–civil-rights positions. With the president traveling around the country campaigning for reelection, Secret Service is taking him very seriously."

Jordan, the civil rights leader, had been shot and nearly killed in Fort Wayne, Indiana, the previous May, and it had brought all the trauma of the 1950s and '60s civil rights movement back to the national stage.

In the Bureau ethos, when you get a call from headquarters, you drop whatever you're doing and get right on it. Technically, the academy is part of Headquarters Division, but in the BSU back then, a request from on high was itself an unusual situation. Normally, they pretty much ignored us, which was fine with me as we tried to develop our program. Until that point, our work typically came from local law enforcement agencies that would request our consultation and send us all relevant case materials: crime scene photos, detective and witness reports, autopsy protocols, lab analysis of physical evidence, that type of thing. From that we would come up with a profile of the type of offender they should be looking for to help focus their investigation. Depending on the circumstances and type of case, we might also come up with proactive strategies to help flush him (and it was almost always a *him*) out or get him to make a move.

This was different. Not only did we have a name at the outset, but the higher-ups at the FBI were the ones requesting our involvement. Unlike all of the requests that had come from local agencies, this was the first case to come from Bureau headquarters itself, important enough that they were willing to take a chance on our still-experimental approach. It was the kind of request that could put us on the map if we were successful. And it was a personal challenge to me to see if we could reverse engineer our analytical process to get our man; and then, if we did, whether we had the goods to properly charge him and bring him to justice.

I CALLED OVER TO THE ACADEMY MOTOR POOL AND ASKED IF THEY HAD A CAR I could have for the afternoon; otherwise I'd have to drive my own car and submit for mileage reimbursement. With all my traveling, I had to fill out enough expense reports as it was, so I'd just as soon not have to do another. The pool did have one available to me, so I hiked over to the garage and signed out one of the standard Bureau Ford sedans. I drove down the winding road that takes you through the marine base that was our landlord, and out onto 95 North. In those days, back in the eighties, it wasn't yet the extended parking lot it's since become at most times of the day, and you could get from Quantico to downtown Washington in less than an hour.

As I drove, I thought of the assignment ahead. This Franklin case, if he turned out to be good for all or most of the crimes he was suspected of, would be on a different order of magnitude than anything we'd worked on to date, both in terms of stakes and public danger. Franklin was operating from a distance against individuals with whom he had no relationship, whom he'd never even met. He was not emotionally involved or personally invested with his victims, factors that were usually important clues for us. And he was highly mobile—he could be anywhere.

Though Dave had given me only the bare outlines of the Franklin story, I knew what was on the line for our unit. Because the profiling program was young and still relatively unproven, this could go a few different ways—most of them bad. While I'd been full-time profiling since January, the BSU didn't have much of a track record within the FBI yet. High-profile cases such as the Atlanta child murders, San Francisco's Trailside Killer, and the brutal murder of Bronx special ed teacher Francine Elveson would come into our orbit within the next year, but in the fall of 1980, the profiling program was still very much an open-ended experiment in search of validation through a larger proof of concept within the Bureau.

We'd had positive results with the cases we got from local police departments or sheriff's offices. In addition, there had been a couple of dramatic successes that had gotten the law enforcement community's attention in the form of what we called 62-D classifications: Domestic Police Cooperation. It meant that even though the FBI didn't have primary jurisdiction in the case, the Bureau might possess some investigative tool or test that could be helpful to a local police agency. Such requests would be handled through the closest field office and were supposed to be channeled through headquarters. More often, it seemed, the special agent in charge of the field office (we call them SACs, with each letter pronounced, "S.A.C.," as opposed to assistant special agents in charge, whom we refer to as ASACs, pronounced "A-sack") would say something like, "Screw headquarters. I want to talk to you." SACs are pretty high up on the Bureau food chain, so they usually got their way. Eventually, each field office would assign a profile coordinator, but that hadn't happened yet.

In December 1979, I got a call from Special Agent Robert Leary of the Rome, Georgia, resident agency. Resident agencies are smaller FBI installations in areas not large enough for a field office. They generally work in concert with the nearest field office. Several days before, a pretty and outgoing twelve-year-old girl named Mary Frances Stoner, a drum majorette in her school's marching band, disappeared just after being let off the school bus at the end of her driveway, which led a hundred yards to her home in nearby Adairsville, in the northwest corner of the state. Her body had been found in the woods, about ten miles away and just over the county line, by a young couple who noticed a yellow coat lying on the ground and went over to investigate. When they saw what was underneath the coat, they contacted police, who came immediately and processed the scene. The medical examiner found evidence of sexual assault, manual strangulation from the rear, and blunt force trauma to the head. There was a bloodstained rock near the body that the police took as evidence.

After reviewing the case file, crime scene photographs, and autopsy protocols, and studying the victimology, I thought I had a pretty good idea of what had happened, and why this low-risk victim in a low-risk setting had been the subject of such a horrific offense.

I gave the police my profile of the UNSUB over the phone, which included an age range of mid- to late twenties; blue-collar occupation, perhaps an electrician or plumber; a military record with a dishonorable or medical discharge; average to above average intelligence but high school education at most; previous sexual assaults, though probably not any murders; probably married because he fancied himself to be a stud, but the marriage would be either dysfunctional or have ended in separation or divorce; drove a black or dark blue vehicle (I had observed that orderly, compulsive people tended to drive darker cars); and a cocky, confident attitude. Since he had to have been around the abduction scene long enough to have noticed Mary Frances, I thought the police had probably questioned him as a potential witness.

When I finished my profile, they told me I had described a guy they had questioned and just released, named Darrell Gene Devier, a twenty-four-year-old white male who had also been a strong suspect in the rape of a thirteen-year-old girl, but had never been charged. He'd dropped out of school after eighth grade despite an IQ listed at 100–110. He already was twice married and divorced, and currently living with his first ex-wife. He'd joined the army after the first divorce but went AWOL and was dishonorably discharged after seven months. He drove a three-year-old, well-maintained, black Ford Pinto. He was a tree trimmer and had been interviewed as a potential witness because he'd been working on trees for the local power company in the vicinity of the crime scene for about two weeks before Mary Frances was abducted. The police had enough suspicion that they'd scheduled him for a polygraph exam that day.

I told them I didn't think that was a good idea. From my experience with the prison interviews, I'd come to the realization that people without a conscience, who can rationalize what they've done through depersonalizing their victims, don't have the same emotional and physical reactions to lie detector tests as ordinary men and women. They called the next day to say that Devier's polygraph was inconclusive, which I thought would only reinforce his own perceived ability to cope with the stress of interrogation.

Now that he knew he could beat the box, there would be only one way to get to him. I told them how to stage the interrogation scene, with the bloodstained rock just in his line of sight but positioned so he'd have to turn his head to look at it. If he is the killer, I told them, he will not be able to ignore the rock. And since Mary Frances had been struck several times in the head, I knew there was a good chance he got blood on him, which we could use to make him unsure of himself. Together with implying that the girl actually seduced him with her attitude, the combined police-FBI team got him to confess, even though Georgia was a death penalty state. This scene was portrayed in the first season of *Mindhunter* on Netflix, and the lesson is: *Everybody has a "rock."* You just have to figure out what it is.

Several months later, in the spring of 1980, I got a call from Police Chief John Reeder of Logan Township, Pennsylvania, who was an FBI National Academy graduate. He and Blair County district attorney Oliver E. Mattas Jr. had been referred through Special Agent Dale Frye of the Bureau's resident agency in Johnstown. About a year before, a twenty-two-year-old woman named Betty Jean Shade had been walking home at about 10:15 P.M. from a babysitting job when she apparently disappeared. Four days later, a man on a nature walk stumbled on her mutilated body in an illegal garbage dump on Wopsononock Mountain, near Altoona. Not only had she been sexually assaulted, the county coroner said it was the "most gruesome" death he had ever encountered.

From the severe facial trauma, it was clear to me that the UNSUB knew the victim well. But unlike some of these I've seen, this was not a sadistic torture killing; most of the wounds were postmortem. Other behavior cues suggested an introverted loner, between seventeen and twenty-five, from a dysfunctional background with a broken family and a domineering mother. I was pretty sure he lived between Betty Jean's home and the body dump site.

In our studies, we'd seen that well-planned crimes that showed a high degree of control and thought—and usually not much in the way of physical evidence left behind—were indicative of one type of offender: organized. By contrast, crimes seemingly committed spontaneously, the victim more one of situation and circumstance than a personal target, and the scenes in a state of disarray, with abundant physical evidence, were the work of a more disorganized subject. The degree of organization or disorganization of the crime was one of the key insights into the personality of the criminal. The Shade scene was what we called a mixed organized-disorganized crime presentation, which can mean a lot of things, but in this case suggested we were looking for more than one offender.

Police had two suspects in mind. One was the victim's live-in boyfriend, Charles F. Soult Jr., known as Butch. The other was the man who found the body, who lived within four blocks of Betty Jean's house and had tried unsuccessfully to pick her up on several occasions. He had a history of antisocial behavior and his alibi was unprovable. But he was married, with two children, and the postmortem mutilation just didn't fit in with this profile. Butch Soult fit more of the profile elements, including a domineering mother and a reputation as being inept with women. His professions of passion for Betty Jean seemed over the top to me. I considered the possibility that they'd had an argument, she'd threatened to leave him, and the situation had escalated from there. When I found out Soult's brother Mike was a trash hauler who'd had the same kind of upbringing as Butch, I became even more interested. I thought we could get the truth through Mike, hammering home that all he did was have sex with Betty Jean when Butch couldn't perform, then help dispose of the body afterward. In the end, Butch was convicted of first-degree murder and Mike, with a plea agreement, was sent to a mental institution. I'm convinced these two would have killed again if not apprehended.

This was the most gratifying part of the job: helping bring

justice to those victims and their families. Because just as these creeps depersonalized their prey, as someone like Darrell Gene Devier did, or carried out their depraved will to viciously punish them, like Butch Soult, my colleagues and I took the exact opposite approach. We personalized and glorified the victims. We had trained ourselves to visualize and actually feel what they went through at the very worst moments of their often all-too-brief lives. I think this was one of the lasting legacies we established. So many of the fictional portrayals of FBI profilers that have come along over the years talk about the importance of being able to put oneself in the mind of the criminal, which we certainly had to be able to do. But being able to put oneself in the mind and shoes of the victim is equally important. It not only helped us contextualize the entire physical and psychosocial environment of the crime—how the victim's reaction affected the offender's attitude, behavior, and actions as the crime unfolded—but also gave us even more motivation to work for the victim, who could no longer speak for her- or himself.

Chief Reeder, who had been trained in our methods, declared publicly that we had been directly instrumental in aiding the investigation into Betty Jean Shade's murder, strategizing the interrogation, and figuring out how to obtain statements from the offenders.

This was the kind of positive publicity we were starting to get for the profiling program. Like John Reeder, a number of police chiefs and sheriffs around the country had called or written to the Bureau saying we had helped narrow a suspect list, focused an investigation, and/or helped with arrest and interrogation strategy. We were even getting calls from SACs saying they had to face the media on a particular case, asking what should they say that would be informative and proactive but not give away details they would later need in order to qualify or disqualify suspects. But, like Stoner and Shade, these had all been local cases that only generated notice within the law enforcement community. We hadn't yet been involved with a case that garnered national media attention, and serial killers weren't a known concept yet in the public imagination.

I can't say that I yet had a vision for what profiling and our behaviorally based criminal investigative analysis approach would become, or how it would continue to grow, but I knew we had something to build on. In my wildest dreams, I envisioned a "flying squad" with its own lab-equipped airplane that could be dispatched on a moment's notice to major crime scenes with a team made up of a crime scene tech, forensic scientist, pathologist, and one or two profiler-investigators, but I knew that was most likely an over-the-rainbow fantasy. There was already an inherent tension between headquarters and Quantico, which many of the bigwigs referred to as "the Country Club" because of our rural location and the perception that we spent a lot of our time just sitting around, contemplating. We, on the other hand, often regarded the denizens of the Hoover Building as a bunch of administrators and bureaucrats who had little idea of what the operational people in the field really did.

So, if Franklin was the case that could put us on the map within the walls of the FBI, it also posed heavy risks for us. The old-school hard-liners who had come up under the late founding director J. Edgar Hoover's iron discipline and "Just the facts, ma'am" crime-solving approach, who kept score strictly by number of arrests, considered profiling and proactive strategies way too touchy-feely and would have been happy to see the whole idea scrapped and send us all back onto the streets. I was foreseeing a time when we'd have other agents doing what I was doing, but failure on a bigger and more prominent stage could put an end to the entire endeavor.

Of course, the biggest concern—much more than my career or even the future of the unit itself—was that this case was a highly visible ticking clock. A possible serial killer and presidential assassin was on the loose. For the sake of all the potential victims out there, we could not afford to fail.

THE J. EDGAR HOOVER BUILDING, WHICH HOUSES FBI NATIONAL HEADQUARTERS, is a massive gray concrete structure in what has been termed brutalist style by architectural critics. It occupies the prime block of downtown Washington, D.C., bounded by Pennsylvania Avenue, E Street, and Ninth and Tenth Streets, one block south of where President Lincoln was assassinated. It had been open for almost exactly five years, since September 1975, and in that time had acquired the reputation as being among the ugliest buildings in the nation's capital, if not *the* ugliest. It is certainly one of the most forbidding, almost inviting you to turn away and walk in the opposite direction. In that, many said, it accurately reflected the personality of the iconic figure for whom it was named.

I pulled into the secure underground parking garage, showed my credentials, and took the elevator up to the executive floor. In those days, you could still see and smell cigarette smoke wafting from some of the offices. I wouldn't have been surprised if a fair number of the desks had bottles of some sort of fortifying refreshment stashed in a bottom drawer.

The federal government had become involved in civil rights in the mid-1960s, when it was nearly impossible to convict white defendants in the South accused of killing African Americans or those who supported their quest for equal rights. The murders of James Chaney, Andrew Goodman, and Michael Schwerner—three young civil rights workers trying to register African Americans to vote in Mississippi during the "Freedom Summer" of 1964—by Ku Klux Klan members and officials from the Neshoba County Sheriff's Office and the Philadelphia, Mississippi, Police Department outraged much of the nation. But the state of Mississippi refused to prosecute. The same year, all-white juries in Mississippi failed twice to convict Byron De La Beckwith of the murder of civil rights activist and NAACP field secretary Medgar Evers as he was pulling into the driveway of his home. Both trials ended with hung juries. The year before, Alabama had refused to prosecute the four FBI-developed KKK suspects in the bombing of the 16th Street Baptist Church in Birmingham that killed four young girls, aged eleven to fourteen.

Because of these and so many other outrageous lapses of justice throughout the burning South of the 1950s and '60s, it was left to the Department of Justice to prosecute these state murders as federal civil rights violations. FBI director J. Edgar Hoover was an unwilling participant, equating the struggle for racial equality with Communist infiltration into the American way of life. (He equated a lot of things with Communist infiltration into the American way of life.) But Attorney General Robert F. Kennedy, Presidents John F. Kennedy and Lyndon B. Johnson, and finally the 1964 Civil Rights Act pushed the Bureau into a more active role in prosecuting civil rights violations, especially when the state wouldn't act.

By the time Joe Schulte and Dave Kohl were involved with the Civil Rights Division, the dynamic had evolved quite a bit, and even southern states were taking a more active stance (though Byron De La Beckwith, for example, was not convicted in a state court until 1994; and the case of the three Freedom Summer martyrs was not reopened by the state until 2005). By 1980, the decision of whether to go first for state or federal prosecution in racially motivated cases that could be legally deemed civil rights violations was largely one of strategy—which case had the strongest evidence and the best chance for conviction.

When I got upstairs, I walked down the hall to Dave Kohl's office. I knocked lightly on his partially opened door and he looked up and gave me his best sardonic smile. "I hope you're ready for this one, John," he said. "Because, you know, if you screw up, Director Webster's already signed your transfer papers to Butte, Montana. All we have to do is file them." In those days, that was the hellhole of the Bureau, where Hoover would send agents he had condemned to purgatory.

"I'm glad there's no pressure," I responded. As usual, Dave's tone was affable and gregarious, but the subject matter, we both knew, was grave.

Dave had collected all the case materials and had them arranged for me in four or five expandable file folders; this was in the pre-computerized-file days. We sat down to exchange ideas about the challenge before us.

"We really haven't had one this mobile before," he said. "Especially in Civil Rights. Most of those guys are one and done. Franklin's never stayed in one place long enough to get him."

Though it still wasn't part of the cultural lexicon, by then we were already using the phrase *serial killer* to reference a predatory offender who killed three or more victims at different times and places, usually with a cooling-off period after each kill. However, the cases we had seen and heard about in the United States and Europe to this point tended to be motivated by perverse sexual gratification—the acting out of the offender's ultimate fantasy. This guy had a different sort of fantasy, and it had to do with ridding the country of people he considered undesirable simply due to their race. He was clearly resourceful, took precautions, and in most instances seemed to plan ahead. Neither Dave nor I would have been surprised to learn that he stalked his victims for days or weeks before each sniper attack. He was not going to be caught by any of the methods that we'd been able to use for men like Devier and Soult.

What the Civil Rights Section was particularly interested in, Dave said, was whether it would be possible to predict where Franklin would land or even gravitate to next, knowing he was now a high-profile fugitive.

"So, what do you think?" Dave asked. "This within your strike zone?"

"Well," I said, wanting to sound confident but give him as straight an answer as I could, "sometimes when the local cops ask for a profile, it's like they think I'm going to be able to give them the UNSUB's name and address. This time, we have the guy's name, so that's a plus. We can work with an actual biography rather than informed speculation."

"And we've got agents seeking out any family members they can find," Dave added.

"That'll be helpful. The challenge is that his victims of preference are such a broad category. But we know he's got to be under stress as a fugitive, so we've got to up the pressure on him, raise his ass-pucker factor. We'll try to key on his strengths and weaknesses and figure out if he has any comfort zones. I'll have more of an idea about that once I've gone through this material."

The pressure would certainly be on Franklin, but just as much on us. The local newspapers and television stations in places like Missouri, Utah, Indiana, Pennsylvania, and Kentucky were already all over the case, and the national media had picked it up. Law enforcement has a complicated relationship with the press. When we want to get out information the public might respond to, like setting up the context in which someone might have seen, overheard, or been told something, or if we're trying to get an UNSUB to react to a particular proactive tactic or strategy, then reporters are our friends and we are very cooperative. When they release details that we'd rather they didn't, or keep up the public pressure in a case we'd rather keep quiet for any number of reasons, they can be a pain in our collective ass. We and they both understand the tension; we've each got our job to do. We need each other, but sometimes those job requirements don't mesh.

This case was a perfect example of the tricky nature of press coverage. Getting out a description of our UNSUB could be critical in terms of generating leads as to his whereabouts. At the same time, as far as FBI brass was concerned, unless we got him quickly, mounting publicity about a resourceful serial killer who'd already slipped through the police's grasp would be nothing more than a distraction and potential embarrassment. And the one cardinal rule of FBI culture, a carryover from early in the ironclad Hoover days, was *Never embarrass the Bureau!*

If I couldn't come up with anything useful—or, worse, if my assessment diverted manpower and resources in the wrong direction; if this exercise screwed up in any possible way—Dave wouldn't be able to protect me, as his not-so-veiled reference to Butte, Montana, suggested. It could even bite him in the ass, since he was the one who recommended that headquarters put its trust in behavioral science (or BS, as they usually called it).

"Good luck, John," Dave said as I was leaving.

Good luck to all of us, I thought.

CHAPTER 2

I carried the files down to the car and headed back to Quantico. Rather than sit in my windowless office with work conversation all around me, my favorite place to go when I was analyzing cases and wanted to concentrate was the top floor of the library, several stories above. For one thing, there was actual natural light from the large windows, and I could look out onto the rolling green Virginia countryside, which to a certain extent counteracted the grim material I was always working on. And there were no telephones or other distractions, so I could concentrate for as long as I wanted to.

I sat alone at a library table that seated eight or ten and spread the case materials in front of me, so I could start connecting dots and establishing relationships in my mind.

I started off with the basic facts—I wanted to get as strong a sense of the individual as I could before delving into the actual case files. Date of birth: April 13, 1950, though some identification listed it as February 9, 1950; probably his way of keeping anyone trying to get to know him off base, like his use of so many aliases. Either way, that made him thirty years of age. Place of birth: Mobile, Alabama; no surprise he was from the South, growing up in a time of intense racial tension. Hair: brown. Eyes: blue. Eagle tattoo on left forearm. Grim Reaper tattoo on right forearm.

The last solid address was in Mobile, from 1977, and his occupation was listed as security guard. Parents were James Clayton Vaughn and Helen Rau. He was previously affiliated with the American Nazi Party in Texas and the Worldwide Church of God in California.

I read more from the first teletype in the files, dated October 2, 1980. Back then, all-points bulletins—APBs—went out by teletype, the same type of system news organizations like the Associated Press (AP) and United Press International (UPI) used to get their stories out to their client newspapers. It was a large, squarish, dark metal box that sat on a table. The paper came out of the top from a roll, and when a message finished printing, you would tear it off. When a particularly important message came in, a bell in the machine would go off, alerting the person attending it to "Rip and read!" The number of rings indicated the importance of the communication. You grew conditioned to react to the sound of that bell. Unlike today's laser printers, when a teletype machine was printing out you could hear the *clack-clack-clack* all the way down the hall. Less urgent messages and reports were sent by airtel, which was essentially first-class mail that you were required to type up and send out the same day. Among the other facts listed in the teletype, I read: "Vaughn allegedly had his name changed in Upper Marlboro, Maryland, to Joseph Paul Franklin and stated that he was having his name changed so he could go to Rhodesia to kill Black people." There was no evidence he'd ever gone to Africa, though.

What immediately struck me as I read through the teletype was under "Physical Characteristics." Aside from the fact that he wore wigs as disguises, which we pretty much knew from piecing together eyewitness reports, he was nearly blind in his right eye, the result of a childhood accident. Accounts varied as to whether the injury was caused by crashing on a bicycle or playing with his brother Gordon with a BB gun. Was his expert marksmanship a means of compensating for this handicap, this perceived inadequacy? I'm frequently asked if there is one event in any violent offender's background that triggers his future path. Did this unfortunate accident become a trigger for Franklin's life of crime? Not likely, though it may have influenced the method he used to commit his violent crimes.

The file listed several of Franklin's earliest crimes, some juvenile: disorderly conduct, a few weapons possessions, and some where the system didn't bother to prosecute him. The first one that seemed particularly relevant took place a few months after he changed his name. On September 8, 1976, he was arrested on charges of assault and battery the previous day against a mixed-race couple whom he followed in his car from near the Kennedy Center in Washington, where they had attended a concert, to about ten miles out in Montgomery County, Maryland. He trapped them on a dead-end street, opened his window, and sprayed them with Mace. The man, Aaron Keith Miles, noted Franklin's license plate number and went to the police. Montgomery County police officers arrested Franklin. He did not appear for his December trial date, thereby forfeiting his bond and having a warrant issued for his arrest. As far as I could tell, this seemed to be Franklin's first violent action against an African American or mixed-race couple, and had I known no more about him, I still would have predicted an escalation in his racial violence. From there, the list of cases shifted to the most recent events—beginning in northern Kentucky a little over two weeks before I'd gotten the call from Dave Kohl.

AT 2:10 IN THE EARLY MORNING OF THURSDAY, SEPTEMBER 25, 1980, POLICE in Florence, Kentucky—just across the Ohio River from Cincinnati and considered part of the greater Cincinnati metropolitan area—received a report of a robbery at the Boron service station at State Route 18 and Interstate 75. The attendant described the getaway vehicle as a silver and maroon 1975 Chrysler with Indiana plates.

Police found the car across the road at Florence's Scottish Inn, and several police cruisers converged on the motel. It was registered to the man in room 137, nineteen-year-old Gary R. Kirk of Dillsboro, Indiana. Officer Dennis Collins knocked on the door and arrested Kirk on suspicion of robbery.

In room 138 next door, one Joseph Paul Franklin was not happy about all the noise and commotion in the middle of the night. He called down to the front desk to complain. He even threatened to check out if the noise continued. Then he went even further. He called the Florence Police Department and complained to the dispatcher, adding that the Chrysler in the parking lot was blocking in his own car. When, after a while, the Chrysler was not moved, he called the police again to complain. He may have called as many as five times, eventually prompting the chief to get on the phone himself and explain that they were investigating a robbery and it had nothing to do with him.

As Officer Collins was getting ready to leave, the motel clerk told him about the man in the room next to Kirk's who had been complaining about his car being blocked in and wanting to know when the police would be gone.

"I went up and talked with [Franklin] and he said he was interested because Kirk had parked his car behind his Camaro," Collins later told a UPI reporter.

Collins then went back to look at the Camaro and spotted a revolver on the front seat. He called in the license plate for a computer check and came up with a hit: the car matched the description of one connected to a double homicide in Salt Lake City, and the description of the individual police were looking for—a slender male Caucasian with a southern accent—corresponded to the man Collins had spoken to in room 138. A team of officers went back to Franklin's room to arrest him. He offered no resistance. They confiscated two rifles and an additional handgun.

As I would see in future cases, it's not unprecedented for a killer to unwittingly bring about his own apprehension, sometimes simply by complaining to the authorities and thereby putting himself on their radar. Franklin's arrest stemming from a noise complaint is similar to that of Dennis Nilsen, the Scottish necrophiliac serial killer and so-called Muswell Hill Murderer. A few years after Franklin's arrest, toward the end of January 1983, Nilsen, a British army veteran, cook, night watchman, and, for a brief period, junior police constable, strangled the last in a series of young men he had lured to his succession of London flats. In his previous apartment on Melrose Avenue in North London, he had buried his victims under the floorboards and eventually taken the bodies out back and burned them. For his last three, after he moved to Cranley Gardens, in London's fashionable Kensington neighborhood, he dissected each one, boiled the heads, hands, and feet in a pot on his kitchen stove to remove identifying characteristics, and then stuffed the cut-up remains down his toilet.

On February 4, Nilsen wrote a note to his landlord on behalf of himself and other tenants in the building, complaining that drains were blocked and backing up. Four days later, at the request of the landlord, the Dyno-Rod emergency plumbing company sent someone out to deal with the problem. Opening up a drain on the side of the building, plumber Michael Cattran discovered a bunch of fleshy globs and small bones clogging up the pipes. Cattran's supervisor came out to take a look and the next morning, upon returning, the two suspicious plumbers contacted the Metropolitan Police. When Detective Chief Inspector Peter Jay and two colleagues went to Nilsen's flat to question him, they immediately smelled rotting flesh. It didn't take long for Nilsen to admit to "twelve or thirteen" murders in addition to the three at Cranley Gardens. He was tried and convicted of multiple murders and sentenced to twenty-five years to life in prison. He died on May 12, 2018, of complications from abdominal surgery.

Now, the similarity between both cases is that the offenders got themselves arrested by complaining, when, if each man was smart, he would have kept as low a profile as possible. What was different between the two was that when Nilsen was asked by detectives why he had committed his murders, he replied, "I'm hoping you will tell me that," and could only offer that he "worshipped the art and the act of death." As would eventually become clear, Joseph Paul Franklin had no question in his mind why he did what he did.

From his room at the Scottish Inn, Franklin was taken to the Florence Police Station, where he was questioned by Detective Jesse Baker of the felony squad. Though his driver's license was in the name of Joseph Paul Franklin, he had what police described as a "multiplicity" of ID cards in other names.

As I read through the transcript of the interview, I saw a suspect who was quite talkative but, shall we say, considerably less than forthcoming. Even after Baker repeatedly warned him that it would be a lot better for Franklin if he said nothing rather than lied, he kept on talking, raising questions he couldn't answer or outright contradicting himself. Finally, he had to admit that his car wasn't even titled or registered in his name. Baker then told him the vehicle was listed as stolen. Nothing Franklin said added up.

The thirty-year-old Franklin drove from place to place every couple of weeks, or even days, could not say what he did in each place other than "trucking around," couldn't explain what he lived on other than odd jobs whose specifics he couldn't remember, and could not explain how he got the money to buy the car he was driving or why it had been identified in a homicide in Salt Lake City. He denied having been in Salt Lake City for at least five or six years, prompting Baker to state, "Now, if you were in Utah and I prove that you were in Utah, then you may not have ever stolen anything from anybody, but you sure lied to me, okay? Now, once I catch you in a lie, then I tend to disbelieve, you know, anything that you've said."

After a while, Franklin conceded that maybe he had driven through recently, and then recalled that he might have spent a few days there. He couldn't explain the "coincidence" of firearms found in his car being of the type that were used in the homicides, or why his car had been identified as the one in the Salt Lake City killings, but he did seem particularly concerned that what the police were really going after was sexual encounters he may or may not have had with underage girls who, he said, looked older than they turned out to be.

Franklin wasn't sure when he'd been where, or where he had slept each night, other than "here and there."

"Most people don't know where they're going, but they know where they've been," Baker observed with exasperation. "You don't seem to know either." The home address Franklin gave in Elsmere, Kentucky, also proved to be false.

The detective conducted a good and professional interrogation, reminding the suspect that he was trying to prove one way or another whether Franklin was telling the truth. He stressed that he was in touch with a number of other police agencies that would soon be able to corroborate or refute what Franklin was telling him. He also said Florence PD was in the process of obtaining a search warrant for Franklin's car.

At some point, Baker left the interrogation room, leaving Franklin with Officer Jim Riley. Sometime after that, there was a knock on the door, Riley opened it, and another officer told him the search warrant had been granted. Just then, Baker returned and was speaking with Riley as they heard a noise and saw Franklin climbing out the window. They chased him, but he was gone. Trained police dogs were able to follow his scent for a few blocks, but then lost it.

Detectives traced back where he had been staying in the area and checked out each location, but Joseph Paul Franklin had disappeared.

CHAPTER 3

It didn't take the detectives in Florence long to reach out to Salt Lake City PD regarding the crime that had originally flagged Franklin's car and resulted in his arrest. Salt Lake City already had three detectives assigned to finding Franklin. At the same time, detectives across the Ohio River in Cincinnati also wanted Franklin so they could question him about the murders of two African American teens the previous June 8.

According to the Salt Lake City Police Department report, around 9:00 P.M. on Wednesday, August 20, Theodore Tracy "Ted" Fields had called his friend and former girlfriend, Karma Ingersol, to ask if she would like to go jogging with him in Liberty Park. She was with her friend Terry Elrod and asked her if she would like to join them. Ted picked up the girls in his car. His friend David Lemar Martin III was with him. Twenty-year-old Ted and eighteen-year-old David were African American. Karma and Terry were both fifteen and white. They parked the car at David's house and started jogging toward Liberty Park.

At around this time, a neighborhood resident named Sefo Manu noticed a dark maroon Chevrolet Camaro driving northbound at a high rate of speed on the west loop of the park. He noticed details of the car, including its red trim lines, rear spoiler, dual exhaust pipes, mag wheels, and tires with raised white lettering, because he owned an older Camaro. He said the driver, who had shoulder-length hair, ran a red light, made a U-turn, and pulled into a field.

Inside the park, the four friends jogged south along the west loop. When they reached an area of tennis courts, the girls stopped to rest. The young men later circled back to join them and the four resumed jogging together. Around ten fifteen, as they crossed a road in a crosswalk, they heard a loud bang, which Karma described as sounding like a firecracker. Terry felt a sudden burning pain in her arm, looked down at it, and saw blood. Then there was a second shot, then two more as David stumbled into Ted's arms. He gasped something like, "Oh my God, Ted, they got me!"

The three others frantically pulled David toward the curb, and Ted shouted for the girls to run. While Ted was trying to pull David all the way back to the curb, two more shots rang out and Ted collapsed onto the pavement.

As it turned out, there were several witnesses to various aspects of the crime. Gary Snow and Mary Biddlecomb were standing outside their apartment building when they saw the four joggers pass by. Snow was backing his car out of the driveway when he heard the first of the shots. He later told police he heard six shots in total. He got out of his car and ran back to his apartment to call the police.

Twelve-year-old Michelle Spicer and her friend Carrie

Beauchaine were looking out a kitchen window when Carrie saw a man in a field rise from a kneeling position with a rifle.

Clarence Albert Levinston Jr. was in his car when he saw David Martin stumble. As soon as he heard the next shot, he determined the direction they were coming from and swung his car onto the crosswalk in front of the joggers to give them some protection.

Gary Spicer came out of his house when he heard the shots and saw a man firing a rifle, then crouching and running to a maroon Camaro parked next to his house. The man opened the trunk and threw the rifle in before driving away. He was able to describe for police the shooter's wide-brimmed hat and waist-length jacket.

Marilyn Diane Wilson was inside a nearby 7-Eleven when she heard the shots and rushed out to the crosswalk, where Levinston was already out of his car trying to help the two men. She rolled the victims onto their backs and detected pulses on both. But in the few minutes it took ambulances to reach the scene, Ted had died. David was still alive but died at the hospital a few hours later.

The morning after the shooting, a crime scene team returned to the cordoned-off area and conducted a thorough search. They photographed and measured tire tracks and recovered six .30-30 bullet casings. They recovered the slugs from the victims and sent them with the casings to the ATF lab in San Francisco, where ballistics expert Ed Peterson determined they had all been fired from the same rifle, possibly a Marlin or Glenfield. Detectives fanned out to local gun shops and scoured classified ads for recent offers. They tested each weapon they could locate, but the ballistics didn't match on any.

Not only was the murder weapon elusive, there was no apparent reason for the shooting. Detective Donald Bell talked to Terry at the hospital that was treating her wound and checked into the background of all four victims. He found nothing. None had ever been in trouble or had associated with suspicious people. They were all upstanding young people. Ted's father was a minister. Terry had been raped months before, and there was speculation that perhaps she was the primary target to keep her from identifying her attacker or testifying against him. But nothing ever came of that.

Could the motive possibly be racial, or was the fact that the two murder victims were Black not significant? Police soon learned that Karma's father, Lee Ingersol, who had driven with his brother Mel to the scene as soon as Karma called him, had been against his daughter dating an African American, but said it was nothing against the race; he just thought the social stigma would make things more difficult for Karma. He volunteered to take a polygraph and it was determined he had nothing to do with the crime or the shooter. Terry's father, Ralph Elrod, was also considered, being a member of a biker gang and rumored to be a Klan sympathizer, but he turned out to have liked Ted and said he had no problem with his daughter dating African Americans. He also passed a polygraph.

African Americans in the area weren't convinced the attack wasn't racially motivated. The Mormon church was the dominant force in Utah society and culture, and there was a long history of antipathy and exclusion of Blacks from the upper reaches of the church. It had only been two years since African American men were allowed to be ordained for the priesthood. In addition, a KKK chapter had recently been established in

Salt Lake City. Police Chief Elbert "Bud" Willoughby tried to reassure the public that the crimes had nothing to do with race. He and Mayor Ted L. Wilson met with James Dooley, NAACP Utah chapter president, to discuss the investigative belief that the crime was not racially motivated. In retrospect, moving the focus away from race just wasted time and resources.

Drug connections and revenge by underworld pimps were also explored—not because the victims were involved in criminal enterprise personally. A close friend of Ted's said the two of them had dated two nineteen-year-old prostitutes and tried to convince them to give up the trade; perhaps the reason David was targeted was because from a distance he looked like Ted's friend. Police questioned the women and their pimps but came up with nothing.

When all the other leads had been exhausted, the police returned to the racial angle. The U.S. Department of Justice assigned Assistant United States Attorney Steve Snarr to oversee the investigation for possible federal civil rights violations. Snarr wasn't sure one way or another. The FBI got into the case through the Salt Lake City field office under the supervision of Special Agent Curtis Jensen. By this time, both the Bureau and the police had concluded the gunman had left the area, so the Salt Lake City field office sent out a teletype notice nationwide on October 2 with details of the case. The same day, the director's office sent out the teletype listing by place and date all the crimes of which Franklin was suspected, which I had read in the file.

One possible lead interested Detective Bell. A college student named Micky McHenry had been trying to make ends meet by working a couple of nights a week as a hooker. The

Sunday night before the shootings, she was sitting on a wall at her usual pickup spot on South State Street when a man in a brown Camaro drove up and asked her to go with him. She told him her rate and then got into the car. He said his name was Joe Hagman. They drove around for a while, stopped to get a sandwich, then drove back to his motel room. In the course of their conversation, he told her he hated Black people and asked if she had ever been with a Black man. He said he didn't approve of white women even speaking to Black men. He added that he was a member of the Ku Klux Klan and, according to a complaint later filed by the FBI at the request of the Justice Department's Civil Rights Division and quoted by the Associated Press, he indicated he had killed Blacks in the past and asked Micky to make him a list of Black pimps in the area so he could kill them.

What particularly interested Bell was that as they drove past Liberty Park and Hagman asked her who hung out there, she told him the east side of the park was mostly white, the west side was mostly Hispanic, and the middle was where the Blacks and pimps hung out.

Hagman showed her two handguns he had in his car. By this time, she was pretty nervous about him and said she didn't want him to kill anyone. When he took her back to his room in the Regal Inn, she noticed two rifles leaning against the wall in the corner. As they were undressing for sex, she noted he bore tattoos of an eagle and the Grim Reaper.

When they were finished, Hagman took McHenry back to her apartment, where she introduced him to her roommate Cindy Taylor. The three of them talked for a while and then he left.

Bell asked McHenry if she could identify the rifles from a

photo array of various weapons, but she wasn't able to. She was able to help police with a composite sketch of Hagman. Several other women who had talked to a man driving a brown or dark maroon Camaro in Salt Lake City, alternately calling himself Joe and Herb, also recognized the sketch. Each one recalled his ranting about African Americans. One of the women mentioned she was a lifeguard at a public pool in Liberty Park. He replied that he wouldn't go because Black people swam there. Two other female hitchhikers he picked up on the afternoon of the shooting remembered him using a racial slur to say that he hated to see white girls with Black men "because it wasn't right," and were then frightened of him until he dropped them off.

If all of these encounters were with the same man, Bell reasoned, then the murders were racial, and this was a very dangerous individual indeed.

CHAPTER 4

In Cincinnati, the deadly June 8 shooting of two African American male teens there had obsessed city homicide detective Thomas Gardner for months. When Franklin was identified as a sniper suspect across the river in Florence, Kentucky, Gardner thought he might have the break he'd been looking for.

Cincinnati cousins Darrell Lane, fourteen, and Dante Evans Brown, thirteen, were shot with a high-powered hunting rifle from the Bond Hill railroad trestle as they walked along Reading Road below on a hot Sunday evening.

"The weapon has been identified as a .44-caliber Magnum carbine," the file stated. This seemed in keeping with Franklin's perceived modus operandi (M.O.), which covers the elements necessary to commit the crime, such as a means to break into a house, bringing a gun to a robbery, or the way an offender lures a victim into his control. The M.O. can evolve as the criminal becomes more experienced and learns what works best. Along with M.O., we consider what we call the offender's "signature," which describes the elements of the crime that satisfy or emotionally fulfill the offender. These could include taking souvenirs, torturing the victim in a particular way, even coming up with a script for the victim to perform during a sexual assault. Unlike M.O., signature doesn't change much, although it can become more elaborate over time. In Franklin's case, shooting victims from far away with a high-powered rifle would be classified as M.O., while selecting African American victims would be signature.

The two young Cincinnati victims, Darrell and Dante, had just left their grandmother's house to go buy candy. Darrell's sister heard the shots and raced out of the house. By the time she reached them, first responders were ministering to the two boys. Darrell's father, a paramedic, was in the first rescue squad unit that arrived on the scene.

But his son had died instantly. Dante was brought to the hospital clinging to life. His mother, Abbie Evans, was attending Darrell's funeral a few days later when she was given the message to get back to the hospital right away to be able to see her beloved middle child alive for the last time.

"It's devastating. It's a void. You never get over it," she told a *USA Today* reporter more than thirty years later.

At the time the two boys were killed, my wife, Pam, and I had two little girls: Erika was five and Lauren was six months. Pam had recently returned from maternity leave to her job as a reading specialist teacher in the Spotsylvania County, Virginia, public school system. I have always tried to put myself mentally and emotionally in the victim's head, as well as that of the killer. But this was just staggering to me, the idea that two innocent children could be taken away from life for no reason on their way to buy candy. It was sickening, and I'd be less than candid if I denied that a lot of people in law enforcement like me have a very hard time giving their kids the freedom and independence they need to grow, seeing all that we've seen.

Likewise, Detective Gardner couldn't fathom why someone would lie in wait to kill two adolescent boys he most likely had never even met. Maybe it was simply a sick thrill killing, but he wouldn't discount the possibility that this was a racial hate crime.

After seeing the director's and Salt Lake City field office teletypes, Detective Gardner got in touch with Salt Lake City PD sergeant Robert Nievaard, and the two men agreed there were similar elements in their two cases that were worth looking into, along with the other possible connections suggested by the two teletypes.

As it turned out, this was just the first of several connections that came about because of the Salt Lake teletype. The two cases certainly seemed to fit in with the M.O. of a sniper-style shooting of a mixed-race couple in Oklahoma City the previous October, a nineteen-year-old African American at an Indianapolis shopping mall in January, and another mixed-race couple in Johnstown, Pennsylvania, in June. If they all did turn out to be linked, then we were dealing with a particularly efficient and deadly serial killer, one who traveled with ease from state to state and never got up close and personal enough with his victims to leave much behavioral or physical evidence, other than the rifle ballistics.

Taken together, all of these incidents described a killing spree that, at a minimum, had been going on for at least a year and probably longer.

The first case under consideration for possible linkage took place in the late afternoon of Sunday, October 21, 1979. As described in an FBI report on a multi-agency intelligence meeting held on October 16, 1980, forty-two-year-old Jesse Eugene Taylor had just left a food store in Oklahoma City and was carrying grocery bags across the parking lot toward his white Ford, in which his common-law wife Marion Vera Bresette, thirty-one, and her three children by a previous marriage, ages nine, ten, and twelve, were waiting. Taylor was Black and Bresette was white. They were coworkers at a nursing home in the city.

As he approached the car, Taylor was struck by three shots, apparently fired from the empty state fairgrounds across the street, about two hundred feet away. He collapsed and Bresette ran around to his side of the car and knelt over him; she was hit in the chest. The children screamed. Both adults died on the spot.

There were several witnesses to the shooting, including sixteen-year-old supermarket employee Charles Hopkins and another shopper, Vince Allen. Police arrived within minutes, dealt with the carnage, questioned the witnesses, and took care of the children, whom they also questioned as delicately as they could. No leads were developed from the eyewitnesses. The bullets recovered from the parking lot were clearly from a high-powered rifle, as were the matching shell casings that were found in a grove of trees across the street.

The Indiana incidents had come next. On January 12, 1980, Lawrence Reese, a lifelong Indianapolis resident, was shot and killed by a bullet fired through the plate-glass window of a Church's Chicken restaurant where he worked, shortly before closing time. The shot was estimated to have come from a rifle with a telescopic sight about 150 yards away.

Just forty-eight hours later, on the night of January 14, a nineteen-year-old Black male named Leo Thomas Watkins was shot and killed, again through a plate-glass window, inside a Qwic Pic Market grocery store at an Indianapolis shopping mall. Leo and his father, Thomas Watkins, were about to begin an exterminating job.

To show how we were still operating largely in the dark and on conjecture, the telegram noted, "Victims in above two incidents were not related." The fact that Reese and Watkins had no known connection was making FBI investigators doubt if the same UNSUB was involved. They looked pretty similar to me, though.

The telegram went on: "Police speculate a Marlin 336 lever-action was utilized." And, by the time I was creating my fugitive assessment, ballistics tests had confirmed that the bullets in the two murders were fired from the same Marlin .30-caliber rifle.

But the Indiana case included in the Bureau file that had generated the most public attention was a shooting that fortunately did not result in a death. It was the event Dave Kohl had mentioned and was listed as an "Ongoing FBI Case."

On May 28, 1980, the forty-four-year-old attorney, civil rights activist, and president of the National Urban League Vernon E. Jordan arrived in Fort Wayne to speak to the league's local chapter fundraising dinner and awards ceremony and checked into the Marriott hotel where the event was to be held. Jordan knew the state well. He had attended DePauw University in Greencastle as an undergraduate, one of the school's few Black students at the time, before earning his law degree at Howard University in Washington.

After a speech to an audience of around four hundred, Jordan went to his ground-floor room to call his wife, Shirley, and then went to the hotel's bar to greet and mingle with some of the dinner attendees. He got to talking to a thirty-six-year-old white woman named Martha Coleman.

When the bar closed and Jordan indicated he would still like some coffee, Coleman offered to drive him to her home in a racially mixed neighborhood in south central Fort Wayne. They drank coffee and talked for about half an hour, then Coleman drove him back to the Marriott. As they stopped at a red light about two and a half miles from the hotel, three white teens in another car shouted racial slurs at them, then sped away as the light changed. Coleman pulled her red Pontiac Grand Prix into the hotel parking lot shortly before 2:00 A.M., driving around to the side entrance near Jordan's room.

They sat talking for somewhere between three and five minutes, and as Jordan got out of the car, he was hit in the back by a bullet that had fragmented when it first hit a chain-link fence around the parking lot. It ripped through the back of his jacket and exited through his chest. The impact lifted him off his feet and hurled him toward the trunk of the car. Through her rearview mirror, Coleman saw Jordan fall to the pavement. She immediately got out of her car and raced inside the hotel to report the shot and tell staff to call for help. As he lay there, never losing consciousness, Jordan later related, the pain was intense, and he was physically aware of the blood leaking out of his body.

There were no independent witnesses to the crime. Coleman had noticed that the teens who had taunted them at the traffic light had then pulled into a fast-food restaurant parking lot, and the FBI agents who investigated the case didn't believe they would have had time to get to the hotel and set up the shot.

The paramedics said the chest exit wound was about the size of a fist. The surgeons at Parkview Hospital who operated on Jordan five times over the next day reported the bullet fragment missed his spine by about a quarter of an inch and easily could have killed him from intestinal damage and blood loss. As it was, he nearly died in the hospital several days later from kidney failure and pneumonia.

Two weeks after the shooting, when Jordan's condition had been stabilized, President Jimmy Carter sent an air force medical evacuation plane to transport Jordan to New York.

With Carter's backing, FBI director William Webster declared the case a civil rights violation and possible conspiracy if more than one individual was involved, bringing the Bureau directly into the case. Their investigation determined that the shot came from the trampled tall grassy area on the far side of an interstate highway off-ramp about sixty yards away, by someone firing from a prone position. An ejected .30-06 cartridge matched the bullet fragments surgically removed from Jordan's body and could have come from about ten different makes of .30-06 rifles.

This sure seemed to match Franklin's presumed M.O., which meant he wasn't afraid to assassinate prominent people. I could see why the Secret Service was nervous and wanted him out of circulation.

The FBI summary stated, "Jordan shot on the night of May 28–29, 1980, from a grassy knoll on an interstate highway while he was exiting a vehicle in a motel parking lot." The phrase "grassy knoll" immediately caught my attention because of its association with the second-shooter conspiracy theory in the John F. Kennedy assassination and reminded me we were on the hunt for another potential presidential assassin. The three components of a successful crime are means, motive, and opportunity. Franklin already had the first two. If he was after President Carter, we had to stop him before he got the third.

The list of possibly linked cases went on. On June 15, 1980, twenty-two-year-old African American Arthur D. Smothers and his sixteen-year-old white girlfriend Kathleen Mikula were walking across the Washington Street Bridge in Johnstown, Pennsylvania, when both were shot from a distance. Smothers was hit first and fell off the sidewalk and into the gutter, shot in the back and groin. Mikula started shouting at passing cars for help. Another shot rang out but missed her and shattered a chunk of concrete on the bridge. Not knowing which way to turn, she stood still for a moment, which gave the shooter time to aim at her again, this time striking her in the chest. Two more shots were fired, one entering her shoulder and traveling down her torso until it lodged in her hip. Smothers and Mikula were both rushed to nearby Lee Hospital. Mikula died during emergency surgery. Smothers died two hours later. None of the people passing in cars could tell police from which direction the shots had been fired. The bullets were believed to have come from a .35-caliber rifle.

The two had been together for about three years and enjoyed popularity among a large group of friends, both Black and white, despite Art's mom, Mary Frances Smothers's concern that Johnstown might not be ready for a mixed couple. Kathy was a gifted artist and Art a talented athlete who wanted to start his own home remodeling business. He ran in the 1978

Boston Marathon, finishing in well under three hours. They were completely committed to each other. At one point, Kathy told Mrs. Smothers she wouldn't want to live without Art.

And once again, the murderer escaped just as efficiently as he had killed.

LOOKING AT THESE POSSIBLY LINKED CRIMES, WHAT IMPRESSED ME WAS THE degree of cooperation between law enforcement agencies. This is not always what happens. Those of us at the academy who used to teach new agents and police fellows used to joke that if you were a killer who wanted to really screw up an investigation, the best thing you could do was drag the body across a county or state line. Fortunately, the investigators in the sniper shootings were a lot more sophisticated.

As investigators in the different states were laying the groundwork for their own cases, Salt Lake City PD continued pursuing local leads. Following up on Micky McHenry's lead, Detective Don Bell found the registration card for Joseph R. Hagman in the motel where she said he was staying. There was no such individual in the FBI databases, confirming Bell's belief that it was a phony name. The FBI lab also searched for latent fingerprints but couldn't find any. This guy had been careful.

The Salt Lake City police investigators visited all of the motels within a thirty-mile radius and examined the guest registration cards. They identified eight that appeared to have similar handwriting characteristics, though all in different names, to the Hagman card. One from the Scenic Motel was from the actual date of the shooting. The Scenic was only nine blocks from Liberty Park.

At one motel, the police were told that a man matching the

UNSUB's description stormed out of the lobby when he saw that the establishment employed African Americans.

At another motel, the Sandman, detectives found a registration card during the proper time frame for a dark brown Camaro. Curiously, there were two license plate numbers listed. It turned out that the first number was the one the guest had listed. But the elderly—and wary—motel owner had a habit of walking the parking lot in the wee hours of the morning and checking plates against what guests had written down. The plate that was actually on the car was from Kentucky: BDC678.

Two Salt Lake City detectives were then able to trace the car's history to a previous owner in Lexington, Kentucky, and contacted the police department there. Though the buyer had given his name as Ed Garland, a Lexington police sketch artist was able to get a good visual description of him from the seller. Cincinnati and Salt Lake City detectives now had a solid car license plate number and several corresponding sketches of the UNSUB.

On September 15, investigators from the Oklahoma City, Indianapolis, Johnstown, Cincinnati, and Salt Lake City PDs convened at the office of the Hamilton County coroner in Cincinnati, presented evidence, and compared their cases. They agreed there wasn't enough evidence to say they all involved the same individual—for example, similar rifles were used in all the cases, but ballistics tests revealed that bullets from at least the Indianapolis and Salt Lake City shootings came from different weapons—but the automobile descriptions were compelling, and the composite sketches looked enough like each other to suggest it might be the same UNSUB.

And ten days after the multi-jurisdictional law enforcement

conference, when Franklin was identified, questioned, and escaped police custody in Florence, Kentucky, the investigators were all pretty sure they had identified their killer.

I mention all of these seemingly mundane details and the tedious shoe-leather police work that went into discovering them because this is how a real and responsible criminal investigation generally unfolds. It's not some hotshot detective cleverly squeezing a confession out of a suspect who slips up in one response at the crucial moment. And it's not a profiler like me looking at crime scene photos and autopsy protocols and magically coming up with the neighborhood and block where the UNSUB lives. It's the meticulous work of analyzing every piece of evidence and following up every possible lead and then working methodically to see how the puzzle pieces fit together or the dots connect. And if people like me and my former unit at the FBI Academy in Quantico can aid in that effort and help the local investigators narrow their suspect list or refine their proactive strategies, then we've done our part of the job.

Federal investigators had also stepped into the search. After the police found weapons in Franklin's car in the Florence motel parking lot, and he escaped from custody, Special Agent Frank Rapier of the Bureau of Alcohol, Tobacco and Firearms began constructing a chart of places Franklin had been witnessed or identified, including motels, restaurants, and places where he'd had his vehicle worked on. The chart included stops in Florence, Atlanta, Birmingham, and Panama City, Florida. As a result, ATF and the FBI were able to secure three federal warrants for Franklin's arrest: for interstate transportation of stolen firearms; purchasing firearms out of state under false names; and unlawful possession of Quaaludes (methaqualone), a prescription sedative that was classified as a Schedule I controlled substance because of its abuse and popularity as a recreational drug.

Cincinnati homicide squad commander Lieutenant Donald Byrd told the *Cincinnati Enquirer* that the escapee had to be considered a serious suspect in the murders of Lane and Brown. A description of Franklin—five feet, eleven inches, 205 pounds, long brown hair, thick prescription eyeglasses, and an eagle tattooed on his right arm—was included in the article. And by this point, Salt Lake City police chief E. L. "Bud" Willoughby, who had initially discounted the shootings in his community as being racially motivated, was now ready to tell the *Enquirer*'s reporters he had learned that though Franklin was normally polite and well-mannered, "he would go into an absolute rage" at the mention of Black people.

This work that the group of local police departments had done establishing the connections between the crimes had been essential to the FBI response that was now in motion. In the following years, establishing the links between crimes would become one of our evaluation tools in the Investigative Support Unit. Though not definitive on its own, we would pose the question: What are the chances that more than one repeat killer is operating in a given area—which could be most or all of the United States—with similar M.O. and similar apparent motivation and signature, such as targeting African Americans, at the same time? The more specific the M.O. and signature, the less likely several cases were merely coincidental and not linked.

One of the things every detective and criminal investigator always has to be careful about is linkage blindness between crimes. Sometimes, you can miss the indicators that two or more cases are related because they took place in different jurisdictions and neither agency reported its case to the others, or because the M.O. wasn't similar enough, the victim profile was different, or any number of other reasons. Linkage blindness can operate in the opposite direction, as well. You can start to link cases that actually weren't committed by the same offender because the motive or victim profile seems to match up, or even because similar weapons were used.

This idea of linkage blindness and M.O. became particularly important as I reviewed Franklin's file, because along with the crimes that appeared connected where local law enforcement had already worked together—Salt Lake City, Cincinnati, Oklahoma City, Indianapolis, and Johnstown, Pennsylvania—there was another open investigation that had been included in his file. It was mentioned as possibly the work of Franklin because it had a racial angle, which was still pretty unusual.

The case involved the murder of four African American males in and around Buffalo, New York. Just three weeks before, on September 22, 1980, a fourteen-year-old boy named Glenn Dunn was shot and killed while sitting in a car in a supermarket parking lot in Buffalo. The next day, thirty-two-year-old Harold Green, an assistant engineer at a local plant, was killed in a Burger King parking lot in the northeast suburb of Cheektowaga. That same night, thirty-year-old Emmanuel Thomas was killed while crossing the street in front of his own house, not far from where Dunn had been gunned down. And the next day, September 24, forty-three-year-old Joseph McCoy was walking down a street in Niagara Falls, about twenty miles away, when he was suddenly struck and killed by two shots.

All four victims had been shot with a .22-caliber sawed-off rifle, leading the media to dub the UNSUB the .22-Caliber Killer. And all four of the victims were African American.

Though we couldn't rule it out, from these details, I was skeptical that Joseph Paul Franklin was the .22-Caliber Killer. Though the motivation appeared similar, Franklin tended not to hang around a location after he had accomplished a sniping kill, whereas the Buffalo killer stayed in the general area. The kills tentatively attributed to Franklin were from farther away, with higher-caliber weapons, which usually were discarded after each crime. The planning had been methodical enough to allow him to slip away unnoticed time after time. The .22-Caliber Killer was more impulsive.

Taken as a whole, the details from Buffalo just didn't seem to match what we knew of Franklin as a killer. But because both Franklin and the .22-Caliber Killer were targeting African Americans with rifles, investigators weren't prepared to rule out the linkage completely. Still, I had my doubts.

As fate would have it, about a week after I was brought into the Franklin case, I was called up to Buffalo to work up a profile of the .22-Caliber Killer, who had possibly killed again. On October 8, two days before Dave Kohl had called me about Franklin, a seventy-one-year-old Buffalo taxi driver named Parler Edwards was found in the trunk of his cab with blunt force trauma to the head and his heart cut out. The next day, the body of another taxi driver, forty-year-old Ernest Jones, was discovered on the bank of the Niagara River, with his throat slashed and his heart also torn out of his chest. His blood-covered cab was found within Buffalo city limits. And the next day, a man roughly matching one possible description of the .22-Caliber

Killer burst into the hospital room in the Erie County Medical Center of thirty-seven-year-old Collin Cole. He shouted a racial epithet and lunged for Cole's throat with a ligature. Only the opportune arrival of a nurse caused the intruder to flee, leaving Cole with serious neck injuries. Edwards, Jones, and Cole were all African American.

"This is the first time that we have had this type of thing on this level," said FBI spokesman Special Agent Otis Cox. "We're looking for either one person or a group of persons with the same types of things in mind."

Thomas Atkins, general counsel for the NAACP, was quoted in the *Washington Post* as wanting to know "whether or not there is some secret and organized effort being made to foment racial strife."

"The rash of murders has fallen like hot embers on a patchwork of school racial battles, cross burnings on lawns and other strife in cities from Boston to the suburbs of Portland, Ore., and from Miami to Richmond, Calif.," the *Post* reported.

FBI director William Webster, with whom I'd had several personal dealings, told reporters in Atlanta, who had started noting Black children disappearing and later found dead beginning the year before, "I think it's a natural temptation born of legitimate fright that a national conspiracy is under way," but didn't think the evidence supported that. I sure hoped not, but this cluster of seemingly racially motivated murders was alarming.

The Buffalo community was in an understandable uproar when Richard Bretzing, the SAC up there, asked me to come from Quantico and work on a profile to see what we could tell about the .22-Caliber Killer. The first thing he wanted to know was whether he was the same UNSUB who had killed the two cabbies, even though the M.O. was different.

It was pretty clear to me that the four shootings from September were all the same offender. They were mission-oriented, assassin-style killings in which the shooter had no relationship with the victims but demonstrated a pathological hatred of African Americans. Within his own prejudiced and delusional system, this was an organized individual who liked firearms. I could see him having joined the military, but he would have soon realized its mission didn't square with his own, and he would have had trouble adjusting to the disciplined military culture. Ballistics tests eventually confirmed that all the victims were killed with the same weapon.

The knife attacks on Edwards and Jones, though, showed much more personal involvement with the victim. They would have taken a longer time and the killer could not quickly run away. Though all of the crimes appeared motivated by racial hatred and fear, for them to have been perpetrated by the same individual with such different M.O.s, I thought, would indicate a pretty severe psychosis, since the sniper crimes were fairly low risk for the shooter while the slashing and evisceration murders were high risk, reflecting rage, overkill, and disorganization.

While it remained unclear whether the .22-Caliber Killer was behind the stabbing murders of Edwards and Jones, by the time I was done working up the profile, none of this matched Franklin, and I was even more convinced that he had nothing to do with these crimes.

Though we wouldn't have it confirmed until a few months later, Joseph Paul Franklin was not the .22-Caliber Killer. In

January 1981, a twenty-five-year-old army private named Joseph Gerard Christopher, who had just been inducted the month before, was arrested at Fort Benning, Georgia, after slashing a Black fellow soldier with a paring knife in an unprovoked attack. A search of Christopher's old house near Buffalo turned up a sawed-off rifle and a large cache of .22-caliber ammunition. He was charged with the Buffalo shootings along with some racially motivated knife slashings in midtown Manhattan the previous December, during a window of time when Christopher was on leave from the army, that earned the killer the title Midtown Slasher. Two other Black victims narrowly escaped being murdered. Interestingly, Captain Matthew Levine, the army psychiatrist who examined Christopher for a possible insanity defense, to whom Christopher confessed that he "had to kill Blacks," said he was amazed how closely the subject fit the .22-Caliber Killer profile I had created.

Ultimately convicted for the Midtown Slasher attacks and the .22-Caliber Killer shootings, Christopher was to serve consecutive sentences for the crimes that exceeded his lifespan. He would end up serving less than thirteen years, succumbing to cancer at age thirty-seven while incarcerated at Attica Correctional Facility in New York State. He remains an interesting psychological case study for us, though, because while his motive was similar in all of his crimes, the variance in M.O. between gun and knife is highly unusual. To this day, it is uncertain whether he committed the murders and eviscerations of Parler Edwards and Ernest Jones.

AS I REVIEWED FRANKLIN'S FILE IN THE LIBRARY IN QUANTICO, THE RESOLUtion to the 22-Caliber case was still years in the future. But because I believed that Franklin was not behind the Buffalo shootings, I was left with a disturbing new realization: in such a short period of time we were seeing two multiple murderers who were motivated not by lust or any kind of sexual perversion, but by pure hate for Black people.

To that point, I'd dealt with crimes of interpersonal violence every day, but they were mostly the result of the sick narcissism of individual human monsters. Though we saw copycats and serial killers who were influenced by other serial killers, there was no danger these crimes were going to spread to a wider group of susceptible people; just as now, while violent video games may stimulate individuals already prone to violence, they are not going to make killers or rapists or carjackers out of ordinary teenage boys or young men who play the games.

As horrible as urban gunmen like David Berkowitz, rapist killers like Ed Kemper and Richard Speck, and sadistic torture murderers like Lawrence Bittaker, Roy Norris, Leonard Lake, and Charles Ng are, there is no chance their perverse designs and deviant psyches are going to achieve larger social purchase and motivate others. We may have a fascination with them and what makes them tick, but that fascination is mixed with revulsion.

With a Joseph Paul Franklin or a .22-Caliber Killer, though, their venomous ideas and increasing victim count are not only imminent dangers in and of themselves, they are the embodiment of a philosophy that actually can draw in and inspire other weak, disenfranchised losers.

That, I think, is one of the main reasons Charles Manson has held such a prominent place for so long in the firmament of American monsters; why he has maintained such a morbid fix on the public imagination. Though he probably shot drug dealer Bernard Crowe and believed he had killed him (Crowe later testified at Manson's trial), Manson never killed anyone himself. What was terrifying about him, though, was his ability to attract seemingly normal, middle-class followers and inspire them to do his murderous bidding without any question of conscience or pang of remorse. This is a power beyond the ability to kill. Even after the trial and his incarceration, he inspired one of his followers, twenty-six-year-old Lynette Alice "Squeaky" Fromme, to attempt to assassinate President Gerald R. Ford in 1975. Seventeen days later, forty-five-year-old Sara Jane Moore also tried to assassinate President Ford. They were the only two known female presidential assassins in American history.

When Bob Ressler and I interviewed Manson at San Quentin, he didn't make a lot of sense with his rantings and ravings against society. But close up, we could see his charismatic dominance and the way he could have exerted control over reasonably intelligent but impressionable people who were looking for some direction and meaning in their lives, and a guru to define it for them.

As I sat in the library staring at the files laid out on the table in front of me, this was what disturbed me so much about Joseph Paul Franklin. Though he didn't have the dark charisma or up-close-and-personal verbal skills of a Charles Manson, his crimes could be just as influential and dangerous—even if only within white supremacist circles.

I had come of age during the civil rights struggles and urban riots of the 1960s, and I saw how they tore apart the country. If we were now on the verge of seeing a new breed of serial killer, one whose motivating energy was nothing more nor less than a hatred of a race that was all too belatedly starting to gain its rightful place in society, then I truly feared for what we in law enforcement, and the nation at large, could be about to face. Vicious and cruel as Franklin was, he represented something far larger and more dangerous than his own miserable existence.

CHAPTER 5

Franklin was now a highly wanted man. Law enforcement personnel throughout the United States would be on the lookout for him. Another teletype went out from the Louisville, Kentucky, field office to the Civil Rights Section at headquarters and the other field offices in areas where Franklin was suspected of having been, describing how he had been positively identified on August 10 and 27, and September 16, as well as purchasing wigs in Johnstown, Pennsylvania, so we knew he was taking precautions in addition to using various IDs in different names. And from his previous experience and track record, we knew he could be anywhere.

In early October, the FBI had issued a federal warrant charging Franklin with unlawful flight to avoid prosecution. A teletype from the director's office to all field offices about Franklin gave his birth name, James Clayton Vaughn Jr., and all of his known aliases: James Cooper, Joseph R. Hagman, William R. Jackson, Joseph R. Hart, Joseph H. Hart, Joseph Hart, Charles Pitts, Ed Garland, and B. Bradley.

After presenting a physical description and the notation that he often wore disguises, the fact that he was blind in his right eye, his known relatives, and his previous affiliations with the American Nazi Party in Texas and the Worldwide Church of God in California, the communication listed several unsolved crimes that were thought to be possibly associated with Franklin. Among them were the incidents in Oklahoma City and Johnstown, and the three shootings in Indiana, including the attempted murder of Vernon Jordan in Fort Wayne on May 29, 1980.

The last phrase of the multi-page teletype was clear and to the point: ARMED AND DANGEROUS.

U.S. magistrate Daniel Alsup issued a federal warrant in Salt Lake City for Franklin's arrest in connection with the shootings in Liberty Park.

With the manhunt a national law enforcement priority, the FBI field offices down south had been working to expand our knowledge of Franklin through interviews with his family members. The next file document I had to work with was a report to the Civil Rights Division from the Mobile, Alabama, field office, detailing an interview with Carolyn Helen Luster of Prattville, Alabama, on October 2. Luster was Franklin's older sister. He also had a younger sister, Marilyn—married name Garzan—and a younger brother, Gordon Vaughn, who, at the time of the Luster interview, had recently been released from state prison in Florida after serving time for a burglary conviction. Gordon had visited Carolyn back in 1973, when she lived in Mobile, and left with her money and jewelry.

Unfortunately, it is not uncommon for brothers brought up in the same dysfunctional, abusive household to end up in similar circumstances; that is, committing crimes and antisocial acts. On the other hand, we see more cases of brothers raised under the same bad circumstances going in opposite ways. For example, Gary Mark Gilmore, the first person executed for murder after the Supreme Court reinstated the death penalty in 1976, has a younger brother named Mikal who became a distinguished music critic and writer.

Another commonality we see—though I would not go so far as to call it a generality—is an overbearing, domineering mother and a weak, uninvolved, or absent father. As we will see in the Vaughn family, the kids got the worst of both worlds.

We are often asked why, in these abusive and dysfunctional families, it is the boys so much more often than the girls who end up as violent criminals. One answer is, they just do. Men are more combative than women, have more difficulty controlling their anger, and are quicker to get into violent confrontations. It may just be that nature created testosterone as a more aggressive hormone back in prehistory when hunting animals larger and fiercer than us was a matter of survival. Second, brutal fathers often take out their own aggression and rage more severely on sons than daughters. And third, we find that women who have been beaten or sexually molested as children tend to be self-abusive and self-punishing rather than lashing out at others, as men do. This could manifest as low self-esteem, substance abuse, prostitution, or unconsciously seeking out brutal or unsuitable men like their fathers in a repetition of their childhoods.

Carolyn Luster said during the time she and Jimmy, as she called him, lived at home together in Mobile, he was an active member of the Ku Klux Klan, becoming involved when he was seventeen or eighteen. The last time she had seen Jimmy was seven years before, in 1973, when he came back home for a visit and found out his mother had died the previous year. Carolyn said he was upset when he learned of her passing. Abused children who hate their parents often have highly conflicted feelings when the objects of their hurt, rage, and resentment are no longer alive to focus on. But Carolyn reported he also became irate when he saw she had a female African American maid working for her, and the argument between Carolyn and Jimmy nearly led her to call the police to get him to leave the house. She told the agents she did not know where Gordon was, and that their sister, Marilyn, might have more information.

Special agents from the Mobile field office then contacted Marilyn Garzan that same day. She traced James's residences from Arlington, Virginia, to Hyattsville, Maryland, to Birmingham, Alabama. She said he had left Birmingham two and a half or three years before and that she had last seen him around that time when she happened to run into him in the Eastdale Mall in Montgomery. She did not know his current whereabouts.

The story Marilyn related of the four Vaughn children's upbringing provided even more insight into the commonalities between Franklin's background and those of the other serial killers I'd encountered. Their father, James Clayton Vaughn Sr., was a butcher born and raised in Mobile, who came back from World War II as a disabled veteran, reportedly having suffered a head wound from enemy fire on Iwo Jima that left him with convulsions, a speech impediment, and need of a cane to walk. His wife, Helen, nine years his senior, was the daughter of Nazi-supporting German immigrants. The family lived a hand-to-mouth existence and the parents fought constantly.

James would beat Helen, at one point causing her to lose a pregnancy. Jimmy was born in a low-income housing project across the street from a Black nightclub. The family moved to Dayton, Ohio, and then New Orleans, and James Senior finally left them when Jimmy was eight, returning infrequently. When he did, he was likely to be drunk and physically abuse the children. Sometimes he used his cane. Carolyn said he was arrested and jailed many times for public drunkenness.

The children got it from both parents, which is fairly unusual. Carolyn said Helen had beaten Jimmy frequently and that he and Gordon were "always in trouble." According to her account, all of the children received regular and severe physical punishment, often for slight or even merely perceived infractions. Carolyn remembered being whipped with a leather belt. Jimmy and Gordon, in turn, took pleasure and satisfaction in abusing cats, such as by hanging them by their tails from clotheslines. Cruelty to animals is one of the surest signs that, without serious intervention, a child may grow up to be antisocial or criminal.

Mobile PD detective Ashbel White, who also was investigating Franklin's background, concluded that the two brothers had essentially been raised more by the two sisters than by either of their parents. Though big for his age and apparently strong, James never went out for any school athletic teams and was considered a loner. None of his teachers seemed to remember him.

Jimmy, Carolyn said, developed a severe hatred of their mother. He was seven years old when she said he suffered the injury that lost him almost all sight in his right eye.

Marilyn said his best friend in high school encouraged him

to join the Nazi party. "He was looking for something," Carolyn added. "Before the Nazi thing, he had attended almost every church in Mobile to listen to the preachers. He was fascinated with religion, searching for the meaning in things." He subscribed to several right-wing and white supremacist publications and took to wearing swastika armbands. He dropped out of high school at age seventeen, and soon after, he met, and two weeks later married, a sixteen-year-old named Bobbie Louise Dorman. With Bobbie looking on, he would frequently stand rigidly in front of a mirror, click his heels together, and practice a Nazi salute.

Ten months later they were divorced, amid claims that he beat her, reprising his father's domestic violence. She listed physical cruelty as the grounds and told authorities that she came to fear for her life. We believe this is more than just "learned behavior" on the part of the abuser. It is compensatory behavior. It represents the psychological need of someone who has been too weak to fend off physical abuse now being strong enough to inflict it on someone else, in this case, a sixteen-year-old wife.

Shortly after they divorced, he moved to Arlington, Virginia, which was then home to George Lincoln Rockwell's American Nazi Party, which he joined.

The FBI summary I was working from stated that Franklin moved to Arlington in 1965, which didn't seem likely since he would have been only fifteen at the time. Later research into his biography by the psychology department of Radford University in Virginia, as well as other sources, pegged his move to 1968, after he had dropped out of high school and after his marriage to Bobbie. In that case, he would not have met Rockwell, who was gunned down in 1967 as he was getting into his car in front of a self-service Laundromat in a shopping center near his Arlington home by John Patler, a former party member whom Rockwell had expelled for what Rockwell termed "Bolshevik leanings." The organization, whose name Rockwell had changed officially in December 1966 to the National Socialist White People's Party though still commonly referred to as the Nazi Party, was taken over by Matthias Koehl Jr. At its height, the party was estimated to have about five hundred members. By the time of Rockwell's murder, it was probably down to about two hundred.

In 1969, after the Manson family murders in Los Angeles of pregnant actress Sharon Tate and six others, Franklin became obsessed with Charles Manson's professed plan for a race war throughout the United States. He was impressed that one leader with a small band of loyal followers could effect such decisive societal action. A confirmation of why Franklin scared and repulsed me as much as Manson did.

Though they grew up in the segregated and discriminatory Jim Crow South, Carolyn said she never noticed Jimmy having a severe hatred for minorities until he moved to Virginia and joined the American Nazi Party. At that point, she said, Adolf Hitler became his hero and he carried around a copy of Hitler's memoir and manifesto, *Mein Kampf*. This may have been so, as Jimmy had little contact with African Americans while he was growing up until he was in high school, when the South was slowly starting to integrate. I wondered if that was when his obsessive hatred of African Americans began. I noted he was fifteen when he stole a copy of Hitler's memoir from the Mobile Public Library and read it for the first time, fascinated by the führer's strength of will and vision of racial purity—proof positive that words have both power and consequences. The fact that he had never knowingly met a Jew didn't seem to matter. The Nazis had been committed to wiping them out before they could put their plan for world domination into effect.

Eventually, James moved to Marietta, Georgia, and obtained his GED (General Education Development, or high school equivalency) degree in December 1974. The following March, he enrolled in DeKalb Community College in Clarkston, Georgia, for a while. By this time, he had also joined the Atlanta chapter of the far-right-wing, white supremacist, and anti-Semitic National States' Rights Party. It was led by Jesse Benjamin "J.B." Stoner Jr., a rabidly racist and segregationist attorney from Georgia who participated in the defense of James Earl Ray for the 1968 assassination of Dr. Martin Luther King Jr. One of Stoner's chief lieutenants was Ray's brother Jerry.

Stoner was a Holocaust denier, who at the same time regretted that it did not happen and was all for bringing back the instruments of control and death all legitimate thinkers knew the Nazis had employed. In 1980, the same year we were looking for Joseph Paul Franklin, Stoner was tried and found guilty for his role in the bombing of Birmingham's Bethel Baptist Church back in 1958, for which he would serve three and a half years of a ten-year sentence. He died in 2005 at age eighty-one, never having renounced his philosophy of hate. Though Stoner only vaguely remembered Franklin when asked about him later, saying he wore thick glasses, this was the kind of leader who fired young James's imagination.

For some of these young men, poorly treated as children

and poorly educated, hate groups like the Klan and Nazis can be appealing. They lend a sense of purpose and mission—however erroneously—to an otherwise aimless life. They suggest strength in a group of like-minded people supposedly working for a common cause. They offer a palatable explanation of why these losers are not getting ahead in life and what unjust forces are holding them back. Perhaps most important, and tied to all the others, is the message that there are groups of people who are inherently inferior to you. Blacks, Jews, immigrants, now Muslims, and, for some, women, are favorite targets, but it can be just about any "other."

James said he wanted to join the Marine Corps but was exempt from military service during the Vietnam years because of his eye injury, and I suspected his obsession with guns was a way of compensating. He briefly joined the Alabama National Guard in 1967, certainly with the intention of feeling more macho and improving his weapons skills. But records showed he was discharged after four months for lack of attendance at drills and being charged with possession of a handgun whose serial number had been filed off. This didn't surprise me. He joined other paramilitary organizations, like the American Nazi Party, which no doubt gave him a feeling of power and belonging, two elements that were severely lacking in his life. I didn't doubt that he got a sexual turn-on as well from that power, just as "Son of Sam" killer David Berkowitz did, even though he never touched his victims. But with Franklin, the gratification of fulfilling the mission would still be the primary turn-on.

There were two other things in Franklin's biographical summary that struck me immediately. The first was the simple fact of his name change. There was the practical consideration of changing his name so he could escape his criminal record and join the Rhodesian army or some other military or paramilitary force. But given the fact that he shared the same name with a father who abused him and that he hated his mother as well, I wasn't surprised that he would be motivated to want to disassociate himself from his previous identity. What was significant to me, though, was his choice of a new name. The "Joseph Paul," according to sources, was derived from Paul Joseph Goebbels, the Nazis' powerful minister of propaganda, who committed suicide with his wife, Magda, after poisoning their six children, the day after Hitler and Eva Braun's suicide in the Führerbunker as the Red Army closed in on Berlin. The last name came from Benjamin Franklin, one of the most prominent of the Founding Fathers of the United States.

These name choices signified to me that this was certainly a mixed-up young man, conflating the values of an American patriot and inventor with a key strategist for the Nazi campaign of hatred, lies, cruelty, and mass murder. While Joseph Goebbels and Benjamin Franklin were both media moguls of their day, what stood out clearly was that these were famous, "important," and influential men—again, something James Clayton Vaughn Jr. was not. It reinforced to me that no matter which of the listed crimes Franklin had committed, he was trying to overcome an overwhelming sense of insignificance and inadequacy. I thought that if we got to talk to him, we'd find out that his personal heroes would be assassins—people like Lee Harvey Oswald and James Earl Ray.

Marilyn Garzan also told the FBI that her brother mentioned he had joined the National States' Rights Party after quitting the American Nazi Party because someone in the Nazi party "was conspiring against him." Now, guys who join these types of organizations tend to be paranoid to a certain extent to begin with. But when you start suspecting your fellow comrades of conspiring against you, you are probably seriously paranoid. This fit in perfectly with what we knew of the typical assassin personality. It helped explain all the aliases and his hostile suspicion of everyone who was different as wanting to take over or diminish him in some way. This meant he would always be hypervigilant, but it also meant we understood one of his emotional hot buttons and might be able to exploit it in some way.

Garzan said she had visited her brother in 1976 in the Washington, D.C., suburb of Hyattsville, Maryland, where he was working as a maintenance man in an office building complex that housed several law firms. He lived in a room in one of the buildings, along with several handguns. She said he seemed to have calmed down somewhat by this time and had dropped out of both the KKK and the States' Rights Party. He said he had left the Klan because of "FBI harassment." Since the Bureau was trying to infiltrate and spy on extremist groups by this point, I couldn't say whether this was his instinctive paranoia, or if he thought he had ever actually come in contact with an FBI informant. But during his time in the Klan, he was exposed to the literature being freely circulated on how to bomb churches and create Molotov cocktails, as well as advanced weapons training.

In any event, he moved back down south, to Birmingham, Alabama, and lived in a rooming house. When Garzan saw him at the mall in 1977, she said, she didn't believe he was working because he had saved a lot of money from his job in Maryland. I found that hard to fathom, given his occupation. If he had money, which he apparently did, I thought he would have gotten it through means other than lawful employment. The two most logical possibilities for where this money was coming from were burglary and robbery.

At the mall, Franklin had told her he had just joined some radical right-wing organization, but she wasn't sure if it was the KKK or not. He had expressed interest to her in rejoining, but she didn't know if he actually had.

Then he told her something that really scared her. He said he had been sitting in his car in the parking lot of an apartment complex and shot a Black man in the chest. He said the police had set up a roadblock, but he had managed to evade it. He wouldn't tell her where or when this had happened, and she didn't know whether it was true or not, but she told the special agents that she thought her brother was quite capable of killing. She was more afraid of her brother now because she had married a man of Hispanic descent and she didn't think he would consider her husband as white. The time she had run into him at the mall, he had pressed her, "You still dating spics?" She said he refused to eat in restaurants that employed African Americans, and that if he ever saw a mixed-race couple out in public, he had no compunction about approaching them and telling them how disgusting it was for them to be together.

While I was surprised he had not gotten into more documented altercations with that kind of brazen offensiveness, it fit the larger portrait that was coming into focus. Looking at the family history alongside the murders he was accused of, I saw an individual who was like a pot of steaming water that had already boiled over. The verbal attacks—and the Mace attack on the interracial couple in Maryland—were just warm-ups for his sniper killings. When he got away with them—remember, though he was arrested for the Mace assault, he didn't show up for trial and so was never punished—he felt emboldened to escalate his action to better satisfy his purpose. Already imbued with a sense of mission that was inextricably tied to his identity and self-worth, once he had experienced the triggering event—his first murder—and realized he had gotten away with that, too, any previous inhibitions he may have had would have evaporated. And after he perpetrated that first killing and experienced the thrill of feeling the ultimate power over his victims' life and death, the assassin within him had been released and the long-held fantasy satisfied. But like every other serial killer, regardless of motive, that satisfaction doesn't last long, and the fantasy must be fed again.

At first his crimes were situational, but as he evolved as a violent criminal, they became more advanced and better planned, lowering the risk for himself and making him even more dangerous. He would keep on killing until he was caught.

CHAPTER 6

Having gone through the case files, the FBI memoranda, news articles (which was a lot more difficult in those pre-internet days), and any additional sources I could access, I spent the day composing my fugitive assessment.

The next day, I sent it off to headquarters, as well as a copy to Roger Depue, the Behavioral Science Unit chief. Roger had recently taken over from the previous chief, who had been an exceptional instructor in the area of practical police problems but wasn't crazy about Criminal Profiling becoming such a popular course and morphing into an operational component of the BSU. Still, he couldn't deny that the positive feedback we were getting both from police departments and field agents around the country as a tool for investigating violent crime was making the whole BSU, as well as the academy itself, look good.

When Roger took over, though, it was a different story. He had a lot of practical experience as a former police chief in Michigan and was 100 percent behind us. He testified effectively before

Congress on the need for more resources to fight the then rising level of violent crime, and we eventually received more personnel as a result. He is both a deeply spiritual man, having spent some time in a seminary, and also the founder of the Academy Group, Inc. consulting service after he retired from the Bureau.

I thought it important to contextualize Franklin's psychic development and motivation both for those hunting him and those who would be dealing with him once he was caught. Even though this was a different type of violent offender than we had previously studied, I felt we could use the same techniques to evaluate and classify him. What I thought would be most important was to try to predict where he might go; that is, to define his comfort zone. Would the publicity and the nationwide manhunt put him under more stress, leading him to grow sloppy and make mistakes, or would it gratify his sense of his own importance? We were clearly hoping to up the ass-pucker factor, but either way, I thought it likely that the knowledge that we were after him would make him both more careful and more erratic. For example, if he'd been making his living by robbing banks, he might be more careful, knowing how heavily surveilled they were, but there were certain things he would have to do, such as find a place to sleep and get some money, and that was where I was hoping his thinking would get more ragged. After all, he had escaped from the Florence police station with nothing, having abandoned his car and whatever he had in the motel room. So, he was essentially starting from scratch. Under the circumstances, I thought he would be like a homing pigeon, going back to places with which he was most familiar. That was the element on which I was betting the heaviest, so I sure hoped I was proved right.

My memorandum to Roger accompanying the document detailed the aim:

> *The personality assessment will hopefully paint a clearer picture of Franklin to arresting agents as well as agents responsible for interrogating Franklin. Furthermore, this assessment is designed to show personality weaknesses as well as strengths of Franklin and to make the future arrest of Franklin as safe as possible for both him and particularly for our Special Agents detailed on this assignment.*

After reviewing his background and upbringing, the assessment explained:

> *Franklin is part of a cycle that continues even now as an adult. He felt his needs, desires and emotions were never really considered; that they were never listened to or considered valuable and important in their own right. Consequently, he felt ineffectual and often worthless. He never had much sense of joy or pleasure in life or any reliable expectation that people would be good to him. He began to think of success only in terms of avoidance of punishment, criticism and ridicule. As a teenager in high school he was delinquent and disruptive. He never graduated from high school although he is average to above average in intelligence. The effects on his psychological and physical development were catastrophic for him. Not only was he abused physically and psychologically, but an early childhood accident left him blind in his right eye. This handicap may have led him to overcompensate for this handicap by*

> *becoming obsessed with weapons as well as firing them with superior accuracy—even with a visual handicap. The effects of his prior experiences as set forth above have created an individual who has a low self-esteem; chronic, low-grade depression; a sense of hopelessness and a lack of basic trust; with corresponding disbelief in the availability of any "helping" authoritarian figure. He believes he should try to solve problems alone rather than share them with others. He does not trust anyone.*

Now, any reasonable and empathetic person merely reading that description would have great compassion for that individual, and I'm no different. When I delve into the childhood and teen years of so many of the people I've studied and hunted—the psychological and sexual abuse, the neglect, the accidents, and the punishment they've endured—I can't help but feel sorry for them and be eternally thankful I had a mother, father, and sister who loved me unconditionally, even when I screwed up, which wasn't all that infrequently.

But though it may explain how they came to be the way they are, nothing excuses the way these violent criminals choose to express their frustrations, their anger, their psychic wounds. Because as we've made clear throughout our writing, unless an individual is so psychotically ill that he is delusional, he can always choose his actions. So, as badly as I felt for what young James Clayton Vaughn Jr. suffered, while I even understood why he had turned to the Ku Klux Klan, the American Nazi Party, and the National States' Rights Party for a feeling of strength, purpose, and belonging, my only aim now was to help put him permanently out of business if I could.

I went on to describe a significant transformation that had clearly taken place in Franklin. He had ultimately left the Nazis and the Klan, not only because he felt both groups were infiltrated by the FBI, but also because he saw them as a group of frequent drunks who only wanted to complain about the Blacks and Jews taking over the country and find camaraderie in their mutual resentment. Franklin, on the other hand, wanted action. He wanted *to take action.*

"What happened to Franklin since leaving high school," I wrote, "is a transformation from a follower in groups with a strong need to belong, to his present need where he now is a leader of his own group—even if it is a group consisting only of himself."

For someone like Franklin, with his background of abuse, neglect, and lack of privilege or adequate education, violence was one of the few ways he could express his resentment and separate himself from all the other malcontents in his various organizations. And when you combine this with his paranoia, his wariness of those around him, then his transformation into the hero-assassin, brave and ready to take on all challenges, is the ultimate expression of self-actualization.

This feeling of wanting to be a leader and in control, coupled with his innate feeling of inadequacy, was demonstrated in his relationships with women. In addition to his short-lived marriage to Bobbie Dorman when she was sixteen and he was eighteen, he had married again in 1979, when he was twenty-nine. Again, his bride, Anita Carden, was sixteen. They had met in a Dairy Deelite ice cream shop in Montgomery, Alabama, in 1978, and Anita gave birth to a daughter on August 25, 1979. By the time we had identified Franklin as a wanted man in the fall of 1980, they were already separated.

As an adult, you do not marry a teenager unless you have a need to control her or him, and/or do not feel adequate in relationships with other adults. In other words, in terms of interpersonal and psychosexual relationships, Franklin had not progressed from his own late teen years. We also had evidence that in between those two marriages he had dated a number of women many years younger than he was.

Not only did he feel threatened by women his own age, he had sexually assaulted an elderly female invalid. This is a clear sign of inadequacy. We see many rapists attack both the elderly and children, not because either is a victim of preference, but merely because they are vulnerable and can't effectively fight back.

We knew that Franklin had become a health food fanatic and an avid bodybuilder and runner. Again, this all had to do with a compensatory building up of his self-image. So, I advised that whoever arrested and interrogated him should pay close attention to his physical condition. If he appeared run down or had noticeably gained or lost weight, he would be more vulnerable. I was hoping that stress would not only lead to a mistake that would help us find him, but also make him more malleable during questioning. I noted:

> *Basically, although Franklin is homicidal, he is equally suicidal. Franklin will appear arrogant, cocky and self-confident, but in reality, he is a coward. His crimes are those of a coward. He ambushes unsuspecting victims rather than killing them directly at close range.*

I warned that psychopathic or antisocial offenders like Franklin often change their M.O. to suit the situation and as they learn from their previous "successes." Since he had been a former security guard, he would know police procedures and I predicted he would probably have a collection of police and security badges and other police paraphernalia in addition to all of his false IDs. His changing of weapons from one crime to another was certainly an indication of a certain type of sophistication about law enforcement.

Despite the indicators that he had spent the past three to five years traveling around the country, I felt certain that with the stress he was now under, he would gravitate back down south, most likely Alabama or the Gulf Coast, to places he felt most comfortable. As distrustful and paranoid as he was, Franklin would have few loyalties. "However," I proposed, "he will be emotionally drawn like a magnet to his young wife and child. They are all he has.

> *Franklin has had few accomplishments in his life. His wife, his daughter and now his mission to eliminate blacks are his only accomplishments.*
>
> *We should expect a wife that is cooperative to a point, however fear of retaliation by her husband will stop her short from providing accurate information relative to the whereabouts of her husband.*

In a summary under the heading of "Franklin's Weaknesses," I wrote:

> *While on the run, he will return to places that he is familiar with. He feels more comfortable in areas where he experienced pleasurable memories in the past. Once again, his wife and other members of his family that he was*

close to as a child will be contacted by Franklin. Returning to these areas is similar to an athletic team having the home court or home field advantage. If an apprehension attempt is made at one of these areas, he will feel better able to meet the challenge.

He would likely still be meticulous in his planning and prepared for a police ambush.

While I expected him to visit Anita, though they were separated, or perhaps his sisters or other relatives, he wouldn't take the chance of sleeping in their homes. I warned that if police located him and attempted a nighttime assault, he was likely to be more familiar with the area and terrain than they were. I thought the best approach, if possible, would be to approach him swiftly by surprise, because homicidal offenders of this type often glory in the fantasy of killing themselves when cornered, or at the moment of greatest drama, or force a "suicide by cop."

I added several pages about interrogation techniques that could be effective if or when he was caught and how police detectives or FBI agents should handle him. I concluded with the offer to discuss any and all aspects of the personality assessment and gave my academy contact information. I was hoping this offer would prove useful later, once Franklin was arrested and agents had a chance to interview him. But first, we had to catch him.

CHAPTER 7

On October 15, 1980, a couple of days after I turned in my assessment, agents in the Mobile field office located and interviewed Franklin's estranged wife Anita.

She was going by the name Anita Carden Cooper because when she met Franklin in Montgomery in 1978, he was calling himself James Anthony Cooper. She told the agents that shortly after they began dating, he left Montgomery for several weeks, and then returned with a large sum of cash. Later, in December 1978, he left again, and after about a week returned with even more cash. After they were married, early in 1979, he left frequently for various periods of time. He never said where he had been or what he had been doing, but he often returned with money.

This fit in perfectly with his earlier claim of having saved a lot of money from his job in suburban Maryland; however, the only two reasonable possibilities for where this money was coming from were burglary and robbery. The amount of cash Anita said he was bringing home seemed to support the theory that he was quite proficient at robbing banks. The previous year, there had been unsolved bank robberies in Montgomery, Louisville, Kansas City, Atlanta, and other places throughout his comfort zone.

In one case, we had a report of a robbery at a branch of the Trust Company Bank on Rockbridge Road in DeKalb County, Georgia, on the morning of Thursday, June 16, 1977. A white male about five feet, eleven inches tall, believed to be in his twenties, wearing a porkpie camouflage hat and green fatigue jacket, entered the building about twenty minutes after it opened, held out a small-caliber pistol, and said to the teller, "Give me the cash or I'll start shooting." No one was injured and the offender escaped with an undisclosed amount of cash. If this was Franklin's work, it was likely his first or one of his first robberies, and the fact that he got away with it so easily would have emboldened him to keep on robbing banks for his livelihood.

Wherever investigators uncovered bank robberies that corresponded to his known or suspected movements, agents were dispatched to interview tellers and show his photo. Several of them identified him, which confirmed for us how he was financing his roaming lifestyle. (We later discovered that he had been inspired to rob banks and develop the necessary skills by reading books about Jesse James and John Dillinger.)

The agents showed Anita photographs of the bank robbery suspect the tellers had identified. She acknowledged them as the man she knew as James Cooper.

There was no reason to suspect that she had any knowledge of his illegal activities or the extent of his racial and anti-Semitic fixation, even though she did have to hear his rantings.

He had told her he was a plumber and led her to believe that his frequent absences were due to complex contracting jobs for which he was paid large sums. Though he sometimes bought her expensive gifts when he was flush with cash, his bank robberies were basically a way to support himself while he pursued his real life's work of killing African Americans and interracial couples, and attempting to foment a national race war.

In July 1979, with their baby due in a month, Franklin told Anita he did not want the responsibility of a baby and was leaving the household, although we know he returned briefly to visit her and his infant daughter in late August before going back on the road. He returned again in October, at which time he was driving a 1972 Plymouth Satellite. He stayed one day and told her he was going to Birmingham.

He came back again in August 1980, this time with the brown Chevy Camaro that police later identified. He told Anita he had been traveling constantly, and that he had been to Canada, Kansas City, and Nevada.

An October 17 teletype from the director's office stated that investigating agents had determined that Franklin had operated for a while under the alias Joseph John Kitts, with a 1951 date of birth, a new Social Security number, and a hospital card from Grady Memorial in Atlanta.

Most important, he had been photographed giving blood for payment at the Montgomery Plasma Center on October 9 and 13, 1980. The center required photographs of all blood donors. The Mobile field office had a copy of the photo—he was not wearing glasses—and they were sending it out to all FBI field offices.

Based on the photograph, Franklin was identified at the

Greyhound Bus Station in Montgomery on Tuesday, October 14, boarding the ten-thirty bus to Atlanta. This was all beginning to fit together.

Within hours of receiving the report from the Mobile field office, FBI headquarters disseminated the information throughout the Bureau. It noted that we didn't think Franklin was aware that agents had identified and spoken with his estranged wife, and that this information should be closely guarded lest he get tipped off. The Fingerprint Division was in the process of comparing known exemplars of his prints with unidentified latent prints from various suspected crime scenes and locations he was thought to have stayed. The Kansas City and Las Vegas field offices were advised to look into Franklin as a suspect for any of their unsolved bank robberies, specifically those from August 1979 through August 1980.

As he traveled, Franklin was going to need money, and we figured that what we now strongly believed was his primary source of revenue, bank robbery, would be too risky. He was experienced enough to know that police departments, sheriff's offices, and FBI agents would all be on the lookout and would harden the targets in locations we thought he might show up. And unless he had an actual death wish, he would also have to believe that having identified him as an armed and dangerous subject, law enforcement officials who spotted him trying to rob a bank would take no chances with their own or the bank employees' and patrons' safety, shooting first and asking questions later. I also thought there was a high chance that, being on the run, he would be too stressed and disorganized in his thinking to carefully plan and execute a bank robbery, which is a high-risk crime under the "best" of circumstances.

When Ted Bundy had felt the stress of being on the run the previous year, we noted that his crimes became sloppier and less well-organized, and his risks were higher. Even someone as arrogant and sure of himself as Bundy was falling apart psychically by the end and showed all the signs of breaking down emotionally. His final crimes—the murder of two young women and assault of two others at the Chi Omega sorority house in Tallahassee, the rape and murder of twelve-year-old Kimberly Leach, and the theft of a van and a car—all showed a psyche in free fall. We didn't think Franklin was nearly as intelligent or socially sophisticated as Bundy and would gravitate to his comfort zone.

The report of the Anita Carden Cooper interview made me even more convinced that he would be in the Gulf Coast region—his comfort zone—though probably not any longer in the Montgomery area, where he had been photographed. He had taken a bus to Atlanta, but from there, he would probably travel even farther south. It would almost definitely be someplace where his southern accent would not arouse suspicion or call attention to itself. And the photo from the Montgomery Plasma Center was an enormous clue.

Though it wasn't nearly the same payday as robbing a bank, when Franklin needed more money, we felt it was likely he would go somewhere to sell his plasma again, as he had twice in less than a week. At least it was legal and wouldn't arouse any suspicion; that is, unless law enforcement authorities could get ahead of him. All field offices were instructed to investigate blood banks that collected plasma and alert them to the possibility that the subject could have visited, or might still stop in. His photograph and description were circulated to the blood banks with the instructions not to confront him if he did show up, but immediately contact law enforcement authorities and FBI headquarters, which would make all resources available, including the fugitive assessment I had prepared.

In my assessment I had referred to Franklin as an assassin-type personality, one "taking cover and placing distance between himself and his victim."

As Dave Kohl had mentioned, President Jimmy Carter, a southern liberal from Georgia, was campaigning throughout the South for the election next month. The threatening letter Franklin had written to then presidential candidate and Georgia governor Carter assailing him for his pro–civil-rights advocacy, which Franklin believed made him a traitor to his region and heritage, was what had put him originally on the Secret Service radar. The letter, signed under his original name J. C. Vaughn, asserted that Carter had "sold out to the Blacks."

We figured no other action Franklin could take would seem more meaningful or fulfilling to his sense of mission and place in history than assassinating the president of the United States. It had been less than twenty years since President Kennedy's assassination in Dallas, and the Secret Service had instituted key lessons learned from that tragic day, including never leaving protectees vulnerable in open cars. But they also knew that another determined would-be assassin shooting with a high-power rifle from a protected perch represented one of the most difficult and challenging threats. In every city or town on President Carter's schedule, Secret Service agents and local police circulated pictures of Franklin with instructions to contact them immediately if anyone thought they spotted him. One of those stops was New Orleans, and that was a large enough city for Franklin to blend in and find the ideal spot for an assassination attempt.

The situation grew even more tense when it was discovered that Franklin had signed in using his original name, James Clayton Vaughn, at the Lighthouse Gospel Mission shelter in downtown Tampa, Florida. He was assigned a bed next to an African American resident and had to listen to a Black preacher at a compulsory after-dinner chapel service. When his sister Marilyn Garzan found out, she speculated that he went there because he figured it was the last place anyone would think to look for him.

Franklin stayed at the mission for three days, just before President Carter was to appear at a rally at Florida Southern College in Lakeland on October 31, about thirty-five miles away on Interstate 4. Florida senator Lawton Chiles, Governor Bob Graham, and former governor Reubin Askew would also be attending. All three were progressives who supported civil rights, so all three were potential targets for a man like Franklin. Philip McNiff, the SAC who headed the FBI's Tampa field office, learned that Franklin had tried to buy a gun in Tampa. Whether it was to rob another bank or try to kill the president or any of the other officials attending the rally, we didn't know.

CHAPTER 8

In the days after I submitted my assessment of Franklin, investigators from more than a dozen law enforcement agencies—including police from Salt Lake City, Cincinnati, Oklahoma City, Johnstown, Indianapolis, Fort Wayne, and Florence, Kentucky, as well as FBI, Secret Service, and ATF agents—converged at Cincinnati's District 4 police headquarters for a two-day "summit" to compare notes on various cases nationwide and coordinate efforts to identify which could be linked to the fugitive. In addition to the sniper attacks, investigators brought details of unsolved bank robberies in their jurisdictions in which the suspect matched Franklin's description. Pooling data allowed a more definite timeline of Franklin's travels to emerge, although nothing of significance was released to the public after the summit, so as not to jeopardize efforts to apprehend—and then prosecute—the subject. In fact, attendees quoted in newspaper articles specifically played down or disputed links to some of the crimes we were investigating Franklin for, including the shooting of Vernon Jordan.

Meanwhile, the FBI was trying to cover as many bases as possible in advance of President Carter's visit to Lakeland. As part of the effort, Special Agent Fernando "Fred" Rivero made the rounds of area blood banks and plasma donation facilities. At about eleven one morning he visited Sera-Tec Biologicals on East Pine Street, a fairly large blood bank that averaged about 120 donors a day. Rivero gave a copy of the Wanted flyer to twenty-five-year-old manager Allen Lee, who said that marginal types, drifters, and even wanted criminals often showed up to make a few quick dollars. The agent impressed on Lee the importance and urgency of the search, with the presidential visit only a few days away. He also told Lee that Franklin was a homicide suspect and "very dangerous."

After Rivero left, Lee mentioned the agent's warning to several of his lab technicians.

As reported by the Associated Press, "At 3 P.M., four hours after the FBI visit, a tired Claudette Mallard looked up from her receptionist's desk to see a 200-pound man walk to the door, wearing brown corduroy pants and a long-sleeved shirt open to the waist. He was carrying a black suitcase."

"Name?" she asked routinely.

"Thomas Alvin Bohnert," he replied, and began filling out a form, giving an out-of-state address as his residence.

He was then examined by Dr. E. C. Wright, sixty-six, who had retired after thirty years as a general practitioner in Waynesville, Ohio, and moved to Florida. Wright went through the standard list of questions about medical conditions and allergies, and screened for infectious diseases like tuberculosis. Wright later said he found the prospective donor strangely quiet and reticent, but his urinalysis was negative, and his pulse and blood pressure were normal. The examination took about eight minutes.

The donation room had twenty-four orange-and-brown leatherette contour beds and on the wall a drawing of Disney's version of the Seven Dwarfs, with Sleepy saying, "No sleeping while donating." Bohnert had an IV inserted in his arm so his blood could be run through a centrifuge with an anticoagulant, the plasma spun out, and then the red blood cell component with added saline returned to his body. The entire procedure usually takes about an hour and fifteen minutes or so.

Two technicians noticed tattoos on the donor's arms: a Grim Reaper on the right forearm and an American eagle on the left. One of the technicians quietly slipped into Allen Lee's office and told the boss that Bohnert seemed to fit the description on the FBI flyer. Allen peered across the donation room. Though the subject's hair was black, not brown as the flyer described it, the tattoos raised enough suspicion in Lee's mind to go back to his office and call the FBI. As it happened, the Bureau had a resident agency in Lakeland, only about five blocks away.

Lee spoke to Special Agent Bruce Dando and explained the situation. "Try to keep him there," Dando urged.

Lee went over to Bohnert's bed and told him he should rest for about fifteen minutes after the transfusion was completed before getting up.

"What if I refuse to stay?" the man asked Lee, but made no move to get up.

Dando immediately called Lakeland PD for assistance. Together with Special Agent Brooke Roberts, Dando met Detectives Gerald Barlow and Ray Talman Jr. outside Sera-Tec and decided to wait for "Bohnert" to leave.

Inside, when the donor was finally told it would be safe for him to stand up and be on his way, he went back to the reception desk, where Claudette Mallard gave him a receipt to sign and then wrote out a check for five dollars.

"Where can I cash this?" he asked her.

She said the banks would be closed by then, but there was a Little Lost Diner around the corner that was open and would cash the check for him. He took the check, picked up the suitcase he had carried in with him, and left the blood bank.

He turned the corner in the direction of the diner, not noticing two cars that were slowly following him. Agent Roberts jumped out of the unmarked Bureau car, flashed his badge, and yelled, "FBI!"

He surrendered without a fight.

Interestingly, though they were taking him into custody, they accompanied him into the diner to cash his check.

Police brought him to Lakeland Police headquarters, where they fingerprinted him and determined that Thomas Alvin Bohnert was, in fact, James Clayton Vaughn Jr./Joseph Paul Franklin. The FBI agents and police officers breathed a huge sigh of relief. Was it just a coincidence that Franklin was in Lakeland, or had he gone there specifically to assassinate President Carter?

Since he denied his identity, it was difficult to know for sure. As he was being held before questioning, the agents noticed he was trying to scrape the tattoos from his arms with his fingernails. But as anyone who has tried to get rid of a tattoo realizes, they don't come off that way; the ink is imbedded in several layers of the skin.

The FBI agents took him to the Tampa field office, where he

was interviewed by Special Agents Robert H. Dwyer and Fred Rivero in a windowless room. Unfortunately, at the time of the interview, the agents were unaware of the personality assessment I had done. Compounding this was what a report from the Tampa field office characterized as "an immediate and massive response from FBIHQ, the media and other agencies having an interest in Franklin, which probably put interviewers under as much tension as they were attempting to induce in Franklin."

The two agents alternated questions about Franklin's whereabouts at the time of each crime he was suspected of. He would not admit any of the crimes, though he freely conceded being a racist and admitted his hatred for African Americans and Jews, which the report characterized as going "much beyond mere bigotry." During the course of the five-hour interview, he was asked if he wanted something to eat or drink and replied that he would like a hamburger, but only if he could be assured it had not been prepared by a Black person, though that wasn't the term he used. One of the agents offered him the hamburger, but he refused it since the agent couldn't guarantee it had not been cooked or touched by any African Americans.

Franklin admitted having been in Salt Lake City from August 15 through 22 and said he had visited Liberty Park but stopped going because of all the racially mixed couples there.

Because the agents didn't have much information beyond the date and place of the crimes, they didn't have much to pressure him with. What they did have was a latent fingerprint that had been recovered from the getaway car in one of the bank robberies, so they tried to press him on that. He began sweating profusely, shaded his eyes with his hand, and looked down at the floor. FBI agents are trained to recognize body language, and Franklin's became highly defensive. Still, he would not admit to anything other than where he had been on certain dates.

After the interview, the U.S. Marshals Service took Franklin to the Hillsborough County Jail in Tampa. There he was told he could make a phone call. He called his estranged wife Anita. The FBI recorded the call. He told her he was being arrested for a number of racially oriented killings. When she asked him the particulars, he told her, "They got me for twelve homicides down here and four bank robberies." Then he added, "And the funny thing is, it's true."

Among the crimes he admitted to her were the sniper murders of the two joggers in Salt Lake City, according to both the Associated Press and United Press International, citing unnamed police sources. The only one he denied to Anita was the shooting of Vernon Jordan. When we analyzed the conversation afterward, we concluded that this denial wasn't because he was ashamed of the crime compared to all the others, but likely due to the fact that, in his mind, his mission to kill the civil rights icon had not been completed.

The next day, Wednesday, October 29, while in a holding cell in the U.S. Marshals office, Franklin confessed some of the murders to Henry Bradford, a federal prisoner in the same cell. When FBI agents contacted Bradford two days later, he told them Franklin had admitted killing Ted Fields and David Martin. This was not uncommon or completely unexpected. Any interrogation is highly stressful for most people, and afterward they need a tension release. We knew from experience that this often comes in talking to or even confessing to someone the accused considers a peer. Another prisoner facing serious charges would certainly fit that bill.

In the Wednesday court arraignment and bail hearing before a United States magistrate, Franklin again denied any killings and claimed the murder charges were trumped up against him because of his avowed racism. “I'm innocent,” the Orlando *Sentinel Star* quoted Franklin telling reporters as he was led in by two FBI agents. He was wearing dark glasses, and a small contingent of African Americans watched from across the street. “They're trying to pin it on me because of my racist views. I'm against race mixing and communism,” he explained.

In what sounded to me like a pretty ironic twist, upon hearing of Franklin's arrest, John Paul Rogers, grand dragon of the Florida chapter of the United Klans of America, was quoted in the *Sentinel Star* as saying that he'd never heard of Franklin, that Franklin was not a member of the Florida KKK (“I doubt he's a member of anything”), and that reports of his membership were just an attempt to smear the Klan's “good name.”

On the other hand, Harold A. Covington in Raleigh, North Carolina, titular head of the National Socialist White Peoples' Party, said to *Los Angeles Times* reporter Jeff Prugh while Franklin was still on the run, “I'm not going to tell you a damned thing that might help get him caught,” and described Franklin as “typical of the decent, white, working people who are fed up with our rotten system.”

Prugh also reached James Clayton Vaughn Sr., remarried after divorcing Franklin's mother, in his Birmingham home. Commenting about the hunt for his son and the murders in Salt Lake City, he said, “It's outrageous. Jimmy wouldn't do anything like that. He's got intelligence. He's been taught better.”

Inside the courthouse, Franklin was charged with the federal crime of violating the civil rights of Theodore Fields and

David Martin III, the two murder victims of the Salt Lake City attack. Citing at least thirteen aliases, numerous attempts at disguise, and "no ties to any community," U.S. Attorney Gary Betz asked that an extremely high bail be set. Betz then went through the list of crimes of which Franklin was suspected, including murders in at least four cities, the wounding and attempted murder of Vernon Jordan, and bank robberies in Tennessee and Georgia. Franklin denied all charges.

He also denied he was in Lakeland because of President Carter's scheduled appearance, replying, "I'm not interested at all in Jimmy Carter."

I didn't put much faith in this statement. I thought it was quite possible that Franklin didn't know Carter was going to be there and Franklin's own arrival in the city was coincidental. But we did know Franklin *was* interested in Jimmy Carter because of the letter he had written regarding the hottest of all of Franklin's hot-button issues. Second, had he known of Carter's upcoming appearance, I believe he would have sensed the same rendezvous with destiny that Lee Harvey Oswald felt when he learned that President Kennedy's motorcade would pass right below his new place of employment.

What greater personal glory could there have been for a man like Franklin than to take out the greatest target an assassin can possibly hope for—the president of the United States—especially when the president symbolized to him a southerner who had turned against his heritage by embracing civil rights and race mixing? Franklin was no martyr; he only would have attempted the crime if he was reasonably sure he could get away with it, and he had already successfully employed the sniper techniques that might have helped him do so. He was highly confident in his own abilities in this realm. And even if he could not have revealed what he had done had he been successful, I believe it would have been the most fulfilling action of his life, because in his own mind, at least, it would have connected him with history, as he believed, like so many other assassins, he deserved.

Though considerations of that sort were important in our ongoing efforts to understand and attempt to predict the behavior of various types of violent criminals, the threat of presidential assassination, thankfully, was now behind us. More pressing was the strategy on how to move forward with Franklin and figure out what we could prosecute him for.

He had denied having committed any murders, bank robberies, or other serious crimes, both when he was interviewed at the FBI office and in open court. Legally, as we all are taught in school, according to the American justice system a defendant is considered innocent unless or until proven guilty. That has a nice and reassuring sound to it, and in fact it is a safeguard against, or at least a discouragement to, capricious or vindictive charging and prosecution. But that doesn't mean that the public or we in the law enforcement community have to assume that we are holding or bringing to trial a presumably innocent man or woman. The very idea is absurd on its face. What the "presumption of innocence" principle—which originates long before British common law, in the ancient Hebrew and Islamic codes—really means is that the prosecution or charging entity holds the entire burden of proving guilt beyond some rigid standard (ours is "beyond a reasonable doubt" for criminal offenses) and that the person charged does not even have to mount a defense if he or she chooses not to. In other words, in court, for the prosecution to win, it must remove all reasonable doubt from every juror's mind that the defendant might not be guilty. I've always considered it significant that a defendant is asked to plead "guilty" or "not guilty," rather than "guilty" or "innocent." In our system, the person charged never has to prove his or her innocence.

So, could we get Franklin to admit anything to an actual law enforcement official that would hold up in court? That was the next challenge.

CHAPTER 9

On November 2, U.S. magistrate Paul Game Jr. in Tampa determined he had seen sufficient evidence to conclude that Joseph Paul Franklin was the man the Salt Lake City Police Department and the FBI believed was responsible for the murders of Fields and Martin, despite his official pleas to the contrary. Among the most compelling elements were the fingerprints on the brown Camaro that had been identified near the Salt Lake City crime scene. Game ordered Franklin's extradition to Utah and granted the government's motion to take handwriting samples from Franklin for comparison with driver's licenses, motel registration cards, and other exemplars with different aliases that investigators believed to have been written by Franklin. If positive, these would help confirm his whereabouts at critical points in the crime timeline.

The same day, Salt Lake County deputy district attorney Robert L. Stott filed a complaint and warrant charging Franklin with two first-degree murder counts. But these were held up while the federal government pursued its civil rights case. The U.S. Marshals Service would be responsible for transporting Franklin from Florida to Utah, accompanied by Special Agent Robert H. Dwyer from the FBI's Tampa field office. This presented us with a possible opportunity: with Franklin handcuffed on a plane for several hours—an unusual and stressful setting, to say the least—he could face an unconventional kind of interrogation, one that might produce the kind of confession that he had avoided previously.

The trip to bring Franklin to Salt Lake City was scheduled for November 8. The Marshals Service had chartered a twin-engine Mitsubishi MU-2 turboprop plane. Along with Franklin and Dwyer, there would be a pilot, a copilot, and three deputy U.S. marshals aboard. In the days leading up to the flight, Dwyer read my assessment of Franklin, and the day before the trip, he called me at Quantico to get advice and strategy on what to do during the flight to get the most out of him.

The private plane rather than seats on a commercial carrier was a great idea, and I advised Dwyer to have the pilots file the longest possible flight plan to keep the plane in the air for as many hours as possible. We knew Franklin wasn't crazy about flying and that he felt uncomfortable whenever he was not in control of the situation. The way he had staged his sniping crimes had proved that trait. Therefore, his stress level would be high to begin with, and he would look for some kind of emotional support from whoever else was on the plane. The only complication was that headquarters had advised the Tampa field office that any conversation regarding the crimes had to be initiated voluntarily by Franklin, and if this happened, it had to be documented with a tape recording.

I thought the best strategy was for Franklin to be accompanied by a senior, authoritative Caucasian agent from the Tampa field office, and Dwyer himself fit the bill perfectly. I suggested he wear the classic FBI "uniform"—crisp white shirt, sharp black or very dark suit, black shoes, the whole works. I wanted him to convey ultimate authority and carry props that suggested we already knew a lot more about Franklin than we were letting on. We knew that Tampa had created a Visual Investigative Analysis (VIA) chart that set out the known activities Franklin had engaged in over the last two years. This would be a perfect prop for Dwyer to have with him, persuading Franklin of the Bureau's high degree of professionalism, suggesting even omniscience.

Dwyer's initial plan had been to attack Franklin's ego and try to intimidate and break him down. I thought intimidation was a good idea, but I thought we should take a different approach to achieve it. I didn't think bringing up the bank robberies again would have much of an impact on Franklin, because this was just his way of making a living, not his raison d'être. I didn't want Dwyer to initiate conversation, which headquarters had already ruled out anyway, but I thought that with the high stress of a long flight in a small plane and the kind of atmosphere we were planning to create, Franklin himself would start talking soon enough.

Once he did, I proposed to Dwyer that he try to lull him into an "us versus them" attitude. I wasn't suggesting he assert that the Nazis or the KKK were good or righteous organizations—Franklin would have seen through that tactic anyway—but that he use what we had already concluded from Franklin's past regarding how he felt about them.

A significant component of behavioral profiling is extrapolation from known facts. We knew that he had joined various hate groups, and that he had subsequently parted company with them. We knew that he had become violent and targeted African Americans, so he hadn't left those groups because of a change in personal philosophy. And we knew he had paranoid tendencies. We also knew from FBI infiltration of extremist groups that a lot of what they did was sit around and talk about their hatred and resentment. Therefore, though we hadn't heard it from Franklin personally (that would come later), it was logical to conclude that the reason he had become a lone wolf, as we called such figures, was because he was fed up with all the talk and no action and/or was fearful of snitches and undercover agents. It would later turn out that we were correct on both counts.

I suggested to Dwyer that he should share the belief that the groups he had once been part of were becoming less and less effective because so many of the members were just talkers or drunks or complainers and weren't really dedicated to any action in furtherance of their aims. That way, even though the agent obviously didn't approve of the murders, he could imply his admiration for Franklin's dedication and sense of mission.

It was the same approach Bob Ressler and I had just used when we interviewed David Berkowitz at Attica State Prison in New York. My dad, Jack Douglas, had been a pressman in New York and then head of the printers' union in Long Island. For the Berkowitz interview, he had supplied me with copies of local tabloids with big headlines about the Son of Sam murders. I held up a copy of the New York *Daily News* and passed it across the table to Berkowitz. I said, "David, a hundred years from now, no one is going to remember Bob Ressler or John Douglas, but they will remember the Son of Sam." I cited the then-current case of the BTK Strangler in Wichita, Kansas, and said that he was writing boastful letters to the police and media and mentioning the Son of Sam in them.

"He wants to be like you because you have this power," I asserted.

I knew that some part of Berkowitz wanted the credit and recognition for his kills, and that was how we initially got him to talk. And the tactic worked. When he went into his much-publicized rendition that it was a three-thousand-year-old demon operating through his neighbor Sam Carr's black Labrador retriever that ordered him to kill, I had heard enough to respond, "Hey, David, knock off the bullshit. The dog had nothing to do with it." He laughed and admitted I was right, that he simply thought it enhanced his story and stature.

I thought the same thing could happen with Franklin if we approached him the right way.

The flight began at 6:00 A.M. from St. Petersburg, which in November meant total darkness. We thought this would add to the spookiness and agitation for Franklin. Dwyer played his part perfectly, arriving at the plane in a three-piece black suit, long-sleeved white shirt, and severe necktie. Early on in our work interviewing serial killers, we had stopped recording our sessions with them when we realized how paranoid they were and how knowing their words were being recorded could inhibit what they told us. But because we knew the tape recorder was required here, we had Dwyer carry it with him to show he was prepared for anything. He also carried copies of the *New York Times* and *Newsweek* reporting on Franklin, and a maroon file folder with the FBI seal on the front. The folder was actually the type handed out to students at the FBI Academy, but it looked very official. Inside, Dwyer had placed several blank sheets of FBI stationary, which also looked very official when the letterhead stuck out of the folder.

Franklin had manacles on his wrists and ankles, which had to make him feel powerless. We would have preferred to have Dwyer and Franklin sitting directly across from each other because I wanted their interaction to seem more like a conversation to Franklin than an interrogation, but the cabin layout made that impossible, so they sat side by side with the tape recorder between them.

It didn't take long for Franklin to start talking. The moment he recognized Dwyer, he mentioned the interview at the Tampa field office several days earlier. Then, as soon as they were airborne, Franklin began asking Dwyer questions about his background and experience. He was impressed when Dwyer told him he had served in the Marine Corps. Franklin brought up *Soldier of Fortune* magazine, and Dwyer said he was familiar with it. That gave the agent the opportunity to mention that he knew a number of people fighting in Rhodesia at the time, which we knew from Franklin's file he had wanted to do but never went through with.

When he noticed the newspaper and magazine articles Dwyer was holding, he asked to read them. Dwyer handed them over. After he read them, as we had hoped, he said he wanted to talk to Dwyer about the crimes he was accused of.

The agent replied that they could do that, but only if Franklin would allow the conversation to be tape-recorded. He agreed. Dwyer then pushed the buttons to activate the recorder and read aloud the "Interrogation: Advice of Rights" form, which Franklin acknowledged. Dwyer asked him to sign it before they continued.

Dwyer then launched into the "Son of Sam" strategy, stroking Franklin's ego by noting the articles were proof that he was a figure of national interest and that he would be influencing a lot of people. Franklin seemed proud and gratified to hear this.

At that point, Dwyer took out the VIA chart. Franklin appeared sincerely impressed that the FBI had taken such an active interest in his travels and activities. As we had discussed, Dwyer didn't bring up the bank robberies and focused on the string of shootings. Franklin acknowledged being in many of the cities where killings had taken place and even expressed familiarity with Liberty Park in Salt Lake City and a fast-food restaurant in Georgia, where another African American was shot and killed. During the course of the conversation, Franklin admitted using numerous aliases, dyeing his hair, buying an assortment of wigs, and even purchasing several vehicles and an array of firearms and bulletproof vests.

The only incident in which Franklin showed little interest was the shooting of Vernon Jordan. Dwyer described his reaction as "flat" and reported that he kept looking out the window, which made Dwyer wonder whether maybe he had nothing to do with that crime.

As with the previous interview in Tampa, Dwyer found Franklin's matter-of-fact racial views almost unreal; he sprinkled racial slurs and abhorrent beliefs into the most seemingly casual of subjects. The hatred seemed to transcend every part of his mind.

Dwyer reported that as Franklin "was a former member of

the States Rights Party, it came as no surprise that he also detested Jews." He was convinced that Jews controlled both the American and Soviet governments. He said he had gone on the FBI tour in Washington twice, and the second time—probably after the tour was moved from the Department of Justice to the J. Edgar Hoover Building—he noticed that the display identified as "the Crime of the Century," the Rosenberg atom bomb espionage case, was missing. He commented that this was because all of the defendants were Jews and Jews had prevented the FBI from remounting the exhibit.

As part of his rant against African Americans, Franklin mentioned that in 1975 he had met a man named Charles who hated African Americans as much as he did, and that while in Franklin's presence, Charles had Maced a Black man he had seen in the company of a white woman. Dwyer had the strong impression that "Charles" was a projection of Franklin himself. Afterward, he checked the files and learned of Franklin's arrest for assault and battery in September 1976 in Montgomery County, Maryland, for the Mace attack on the interracial couple.

There are several reasons a suspect might project his actions onto another persona. The most obvious one is to establish an insanity defense based on multiple personality disorder. This tactic seldom works, but most defendants don't know that. Another reason is for psychological face-saving. In 1985, after thirty-six-year-old Larry Gene Bell was arrested for the abduction and murder of seventeen-year-old Shari Faye Smith in Lexington County, South Carolina, I was asked to interrogate him to see if I could get a confession. After setting up for him the proposition that some people end up doing things as if in a nightmare and finessing him into a reaction to the crime, I asked, "Larry, as you're sitting here now, did you do this thing? Could you have done it?"

He then looked up at me with tears in his eyes and said, "All I know is that the Larry Gene Bell sitting here couldn't have done this. But the bad Larry Gene Bell could have."

In Franklin's case, this conversational tactic occurred not long after his arrest, when the idea of long-term incarceration was probably starting to set in. He would have started thinking about the racial animosity he would face from Black inmates if they knew the details of his crimes, and this could have been a feeble self-preserving attempt to separate himself from his own racist image. It quickly became clear he couldn't sustain this, though. An equally plausible, though opposite, explanation is that he didn't consider the crime in question up to his "standards" as a killer.

Though nearly everything out of Franklin's mouth was suggestive on one level or another, the ego-stroking technique was not producing any direct confessions. Dwyer decided to take the Son of Sam strategy one step further. He said he knew that all of the traveling Franklin had done in the past several years, all the banks he had robbed, and all the sniper shootings he had performed were part of his "historic mission," which Dwyer said he understood to be to kill Blacks and Jews. He suggested that the mission would not achieve the significance Franklin felt it deserved unless it was all written down somewhere, and that the best way to do this was for Franklin to write it all out in his own handwriting on the official FBI stationery Dwyer was carrying. That way, it would become a historical document, the agent stated, akin to what was held and exhibited in the National Archives building in Washington.

Franklin thought long and hard about this offer and Dwyer reported that he was severely tempted to take him up on it, and it was only with the most disciplined restraint that Franklin decided not to.

They had been in the air about seven hours at this point, and Dwyer was feeling emotionally drained from being in the close presence of a man like Franklin and trying so hard to come up with confession strategies. But he also sensed they were nearing a now-or-never moment, and he had to try everything he could think of.

Since Franklin was subject to both state and federal charges, Dwyer said that if Franklin would plead to a federal crime, he could probably work out a deal to be placed in a prison located close to his wife and little girl. If he was found guilty in any of the possible state courts in which he would be tried, the feds would have no influence over where he ended up. Once again, Franklin seemed intrigued by the idea, but pulled back without implicating himself.

With no prompting from Dwyer, Franklin commented that he admired Fred Cowan and what he had accomplished. Dwyer asked who Cowan was and what he had done. Franklin related that Cowan was a member of the National States' Rights Party and had swastikas tattooed on his arm. Several years earlier, according to Franklin, Cowan went to the warehouse where he worked in New Rochelle, New York, heavily armed, and killed four Black people who worked there. He tried to kill the Jewish business owner as well, but he'd hidden under a desk. When the "pigs" arrived at the scene, Cowan killed one of them and then committed suicide.

The actual story wasn't all that far from the one Franklin

told, but with a somewhat different emphasis. At 7:45 A.M. on St. Valentine's Day, February 14, 1977, thirty-three-year-old Frederick William Cowan, a twice court-marshaled ex-GI, bodybuilder, and Hitler admirer who lived with his parents and collected Nazi paraphernalia, arrived at the Neptune Worldwide Moving Company, looking for Norman Bing, a supervisor who was Jewish and who had suspended him for being rude to customers. Unlike what Franklin had said, Bing was not the owner of the company. As Cowan passed through the lobby and cafeteria and toward the office area, he shot and killed three Black employees and a dark-skinned electrician from India. Bing saw Cowan entering the building, left his office, and hid under a table in another room.

Within ten minutes, police had arrived and charged into the building. Cowan shot and killed the first one in and wounded three others. Before long, the building was surrounded by three hundred officers and agents from state, local, and nearby police departments and the FBI, and helicopters circled overhead. They waited out the siege for several hours because of the fear of harm to the hostages and the explosives Cowan threatened to detonate. When they finally went in, they found Cowan dead in a second-floor room of a single gunshot wound to the head. He had not been holding any of the fourteen remaining and hidden employees as hostages.

Technically, Cowan was a mass murderer, while Franklin was a serial killer. But this was just the kind of person we could imagine Franklin admiring and drawing strength from. Although Cowan's crime was basically an attempted revenge killing, Franklin would see a couple of tropes to which he could relate. First, the perceived "owner" of the business was

Jewish, and going after him was a justified way of getting back at the Jews. Second, Cowan had tried to kill as many Black people as he could, and if one of them was Indian, well, close enough. The interpretation for Franklin would be that Cowan acted on his beliefs rather than just talking about them, and that he fulfilled his mission, even at the cost of his own life. Franklin, on the other hand, for all his blustering, had no intention of dying for his beliefs.

Unlike in the first FBI interview with Franklin, in which the interrogating agents alternated questions, I had suggested to Dwyer that whenever there was a lull in the conversation, he shouldn't try to fill it. As part of his insecurity and need to feel in control, Franklin would be compelled to talk. It was a common phenomenon we had seen both during hostage negotiations and in our prison interviews. We called it a "voice vacuum," and the other party often would feel a need to fill it. And this is what Dwyer said kept happening.

As the plane approached Salt Lake City, part of our strategy was to circle over the Utah State Prison at Draper. From the air, it is a gray and white compound of institutional-looking buildings that might even appear to be a factory complex. Seizing the opportunity, Dwyer pointed out the facility and commented that it was where two-time killer Gary Gilmore had been executed by firing squad about three and a half years earlier. He described how Gilmore was strapped to a chair with a wall of sandbags stacked behind him and a paper target pinned to the left center of his chest. Five local police officers stood behind a curtain with small holes through which they aimed their rifles. The medical examiner's autopsy, Dwyer went on, showed that the bullets had completely pulverized Gilmore's heart. That was how Franklin would die if Utah had its way with him, the agent noted. Actually, Gilmore was given the choice between firing squad and hanging, but the drama of the description had the desired effect on Franklin. His attention was riveted on the scene below the plane's window. I thought Dwyer's ad lib was brilliant.

Though Franklin hadn't yet admitted to any murders, I thought Dwyer's first-rate performance during the flight would still yield positive results down the line.

Near the end of his detailed report, Dwyer wrote:

> *In conclusion, it is felt that the many suggestions made by BSU preparatory to conducting this interview were extremely valuable and virtually each technique suggested succeeded to one extent or another. Few interviews are conducted under ideal conditions and investigations seldom have the opportunity to make the preparations that led up to this interview. However, the foregoing narration clearly indicates that the services of BSU are a valuable weapon in the Bureau's investigative arsenal.*

Most of us are attracted to the image of the G-man, and it's a prime reason many of us wanted to go into this line of work in the first place. But let's be honest enough to acknowledge that on one level, the FBI is a government *bureau*cracy like any other, with competing interests, agendas, and power centers. So, that vote of confidence at this still-early and tentative stage of our evolution meant a lot to us in establishing our legitimacy and supporting the development of criminal profiling, investigative analysis, and proactive behavioral strategies that would become the essential core of my work and that of the colleagues who would join me. The proof of concept of our research and methods was coming into focus.

And, in upping Franklin's ass-pucker factor, we had considerably reduced our own.

CHAPTER 10

Because of Franklin's notoriety, the marshals were anxious that he be transferred to the Salt Lake City Jail with as little publicity and fanfare as possible. The chartered plane landed at Salt Lake City International Airport and taxied to the far side of the airfield, where the Utah National Guard had a hangar. But as soon as the plane came to a stop, the marshals looked out the window and saw a television crew waiting on the runway and a Channel 2 helicopter swooping down for a better view. Apparently, someone in local government wanted a media show. The marshals ordered the plane into the cavernous hangar and a tractor pulled the outer door shut. The interior was lined with air police officers carrying M16 rifles.

A deputy U.S. marshal fitted Franklin with a bulletproof vest and put him in one of three windowless vans in a convoy that was led by a V-formation of five Salt Lake City motorcycle cops, an air force truck with a mounted machine gun, and city police cruisers with their sirens blaring. Dwyer commented that the hoopla rivaled a Macy's Thanksgiving Day parade. The plan was for Franklin to be deposited in the basement garage of the jail. Dwyer asked the police official accompanying him what to expect when they arrived. He was told that just about every media person in Salt Lake City would be there. Dwyer immediately worried about a scene similar to when Lee Harvey Oswald was shot before a gaggle of cameras and reporters in the basement garage of Dallas police headquarters.

Franklin had been calm and fairly talkative during the ride into town, but as soon as they reached the garage and he was escorted out of the van, Franklin and his minders were blinded by a sea of camera flashes. The prisoner, wearing dark glasses, lost it at that moment and began screaming that he was being persecuted for his racist beliefs and that "the Communist federal government is trying to frame me!"

Based on our strategy of trying to place Franklin under maximum stress, especially the mention near the end of the flight as to how he could face the same fate as Gary Gilmore at the Draper penitentiary, Dwyer told jail officials he expected Franklin to talk about his crimes in some form of confession to other prisoners as a stress-relieving action. We'd already seen Franklin do this after his first interview with Dwyer and Fred Rivero in Tampa. Dwyer suggested they debrief any inmates who had any kind of contact with Franklin over the next twenty-four hours and see if he said anything to them. Within the time frame, as Dwyer predicted, Franklin did confess a number of his crimes to several inmates.

With a character like Franklin, you can never be sure whether confessions such as these are legitimate or merely braggadocio to other criminals to enhance his image. In and of themselves, these confessions wouldn't be much good in court. But they did give both federal and state prosecutors added confidence that they could proceed to trial.

Though the Salt Lake County attorney's office and Salt Lake City PD detectives felt they had now amassed sufficient evidence to go forward with state charges, they agreed that the best strategy was to go first with the federal civil rights case. The county attorney filed two counts of first-degree murder, to be pursued after the federal prosecution.

On Monday, November 10, Franklin, handcuffed, in leg irons, and accompanied by his court-appointed attorney Stephen McCaughey, was arraigned before U.S. magistrate Daniel Alsup, who had issued the original warrant for Franklin's arrest in connection with the Liberty Park murders. Four armed guards stood outside the magistrate's chambers. When Alsup asked for his plea to the first federal civil rights charge against him, Franklin replied, "Definitely not guilty." In response to the second charge he said, "Same thing." Alsup continued his bail at one million dollars. County Attorney Theodore L. "Ted" Cannon confirmed that the state would defer its own prosecution until the federal case was completed, and U.S. Attorney Ronald Rencher said he did not believe the federal and state cases constituted double jeopardy because the charges were different.

"The federal charges are a separate and distinct matter from the local murder charges," Cannon stated. "The federal complaint supports the theory that the individual so charged did violate the civil rights of both Martin and Fields because of their race and color."

Franklin's arraignment made national news, and one of the people who caught the report was Lee Lankford, a captain in the Richmond Heights Police Department who, as a detective sergeant, had investigated the 1977 Brith Sholom Kneseth Israel synagogue shooting. When he happened to see the television coverage of Franklin's court appearance, he became convinced that Franklin was good for the Brith Sholom shooting. The Salt Lake City crime was close enough to the Richmond Heights M.O., and Franklin looked enough like the composite drawing of the man leaving the synagogue scene. Learning that he hated African Americans and Jews cinched it. Lankford delved back into it, but the case was not yet strong enough to bring to the district attorney. Still, the detective pledged to Gerald Gordon's mother that they would find her son's killer.

In addition to the state murder charges in Utah, first-degree murder charges were filed in Oklahoma County District Court alleging that Franklin had killed Jesse Taylor and Marion Bresette as the mixed-race couple left the supermarket on October 21, 1979. "Franklin told friends and some of his cellmates that he, in fact, committed the homicides here in Oklahoma City and he told them specific details about the homicides," Oklahoma City homicide detective Bill Lewis said to an AP reporter.

John D. Tinder, chief trial deputy for the Marion County, Indiana, prosecutor's office, said following a conference with FBI officials that Franklin was a "strong suspect" in the murders of Lawrence Reese and Leo Thomas Watkins, both shot sniper-style through plate-glass windows, two days apart in January 1980.

The Department of Justice announced it was still investigating Franklin in connection with the wounding of Vernon Jordan.

Officials in Cincinnati and Johnstown, Pennsylvania, also considered Franklin a prime suspect for the June murders in their cities the previous year. It seemed just about every case we'd looked at in connection with him, other than the .22-Caliber Killer, he could be good for.

On Tuesday, November 25, while awaiting trial, Franklin gave a telephone interview to the *Cincinnati Enquirer*, which was interested because of the pending charges against him for the June 8 sniper murders of the two boys, Darrell Lane and Dante Evans Brown.

In describing his escape from the police station in Florence, Kentucky, Franklin offered, "The Lord didn't think my time had come to get caught," explaining, "I was handcuffed to a chair. I prayed to the Lord. An hour later, this blond guy took the handcuffs and left the room. I already knew where the window was because earlier that night a guy had rapped at the window and wanted to know how to get into the Florence police headquarters and I told him." He said that once he had jumped out the window, he ran to the street and was picked up in a car driven by a "high school kid." The driver dropped him off in northern Kentucky; he hitched another ride to Cincinnati, then took a bus to Columbus. The odyssey took him to Charleston, West Virginia; Winston-Salem, North Carolina; and Atlanta, before he made his way down to Florida.

He said he thought he was being held and charged because of the letter he had written to candidate Jimmy Carter in 1976, but again asserted that he had no interest in President Carter as a target.

Franklin claimed he was innocent of all the sniper slayings and would only kill in self-defense. He added cryptically, "I did some stuff. I did a few things. I'm not totally a good guy, you know. But the end justifies the means."

He also granted an interview to KALL radio reporters Mike Watkiss and Dave Gonzales in which he said that "race mixing is a sin against God and nature," and that while he was innocent, whoever had killed the joggers had committed "justifiable homicide."

In December, as the trial date was approaching, federal prosecutors sought a psychiatric examination of Franklin. Assistant U.S. Attorney Steven W. Snarr filed a motion for the examination to see if Franklin "may have been insane" or suffering from mental illness.

This idea comes up all the time in my work and I always have to explain it. The average person cannot see how someone could kill in cold blood, with preplanning, without being insane. I will grant that just about all of the violent killers I have encountered over the years have been mentally ill to one degree or another. They tend to be narcissistic, paranoid, and either completely devoid of empathy or extremely selective about whom they extend their empathy to. We call that type of mental illness a character defect, a term that speaks for itself. But insanity is a legal term, going back centuries in British common law. Though the definition of insanity has evolved over the years since 1843, when Daniel M'Naghten was tried in London for attempting to assassinate Prime Minister Robert Peel and successfully murdering his private secretary Edward Drummond, the basic concept has remained the same as spelled out in what came to be known as the M'Naghten (sometimes rendered McNaughton) Rule:

> To establish a defense on the ground of insanity, it must be clearly proved that, at the time of the commit-

> ting of the act, the party accused was laboring under such a defect of reason, from disease of the mind, as not to know the nature and quality of the act he was doing; or if he did know it, that he did not know he was doing what was wrong.

Practically speaking, this means that if the accused comprehends the distinction between right and wrong and is able to conform his behavior to the rules of society, then he is morally and legally culpable for the crime. So, who would be covered by insanity rules? Someone who was truly delusional and psychically detached from reality. M'Naghten was found to be suffering from severe delusions of persecution, was found not guilty by reason of insanity, and was transferred from Newgate Prison to the State Criminal Lunatic Asylum at Bethlehem Hospital, commonly known by its nickname, Bedlam. I can think of only a handful of killers I would consider legally insane, and Joseph Paul Franklin was not among them.

The judge agreed.

By the time the trial in federal court began, in February 1981, a nineteen-year-old inmate named Robert Lee Herrera, who was being held in the Salt Lake jail on a burglary conviction, got word to the local FBI field office that Franklin had confessed the Fields-Martin and Lane-Brown murders to him after Herrera told him that he had been moved to Franklin's cell block after he got into a fight with a Black man. Herrera said Franklin told him he would kill anyone of any age as long as they were Black, particularly if they were mixing with white. Black men with white women in public, he believed, should die. He reportedly admitted the Vernon Jordan shooting for the first time, but perhaps most interestingly, he confessed to an unsolved crime that no one, up until that point, had been able to pin on him.

In March 1978, *Hustler* magazine publisher Larry Flynt was on trial in Lawrenceville, Georgia, for obscenity charges. Near the Gwinnett County Courthouse, where the trial was taking place, Flynt was shot and permanently paralyzed from the waist down. Flynt's attorney, Gene Reeves, was also seriously wounded, requiring twenty days in the intensive care unit of Button Gwinnett Hospital, but he eventually made a full recovery.

For the nearly three years since, the crime had remained unsolved, and though it was a sniper-style shooting, it hadn't been considered in connection with Franklin's alleged spree. When he was captured in Florida and the extent of his alleged crimes became known, local Georgia authorities said Franklin was "wanted for questioning," but that was as far as any connection had gone. But speaking to Herrera, Franklin confessed to Flynt's shooting, explaining that he'd done it because he was disgusted by the porn publication's occasional pictorials of mixed-race couples.

The FBI agents who met with Herrera conveyed the information to the U.S. Attorney's Office. Several assistant attorneys then interviewed Herrera themselves and had him sit for a polygraph exam, which he passed. Still, any jailhouse snitch is questionable because they all have something to gain from cooperating with the authorities. Though Herrera offered a fair degree of detail about the aftermath of some of the crimes, the attorneys weren't sure any of it wasn't available from previously published sources. And even if he was telling the truth, maybe Franklin wasn't.

As the trial date neared, the prosecutors learned that Franklin had also shot his mouth off to another jail inmate, Richard Hawley, confessing to the Salt Lake City murders and that of a mixed-race couple in Oklahoma City in 1979. Hawley was being held on federal charges of conspiracy to transport explosives across state lines and wanted to talk to the feds about favorable treatment before he had to plead, but the information he provided built on a pattern of Franklin's confessions.

The Salt Lake City trial began on Monday, February 23, 1981, with Judge Bruce S. Jenkins presiding and Assistant U.S. Attorney Steven W. Snarr heading up the prosecution. Demonstrating that Franklin was an avowed racist was not difficult; he had freely admitted it to the press and numerous individuals. In fact, the defense implied that his racism was the reason the government was prosecuting him, rather than that he had actually murdered anyone. The prosecution spent a week on its case, including circumstantial evidence regarding Franklin's vehicle and the witnessed getaway car, tire tread matchups, and sixty-five witnesses. None, however, had seen Franklin fire the rifle, nor was the rifle located.

But they did have testimony from Herrera and Hawley, as well as Anita Cooper, all of whom repeated their accounts of how Franklin had admitted the murders to them. Hawley mentioned the "joy" Franklin had described in killing the two men. Hawley acknowledged that he had pled guilty about a week and a half before but that there was no agreement with prosecutors about how his testimony would affect his upcoming sentencing.

As Hawley was completing his testimony, Franklin shouted out from the defense table, "How long did it take you to make that up, Hawley, you liar!" Even after a warning from Judge Jenkins,

Franklin kept spouting about liars and snitches when prosecutors would mention Herrera, Hawley, and Cooper to the jury.

Among the witnesses called to testify were the two teenage girls who jogged with the victims. They described their terror as Martin and Fields were hit.

Also, the *Salt Lake City Tribune* reported:

> *Three witnesses, Leon Beauchaine, his daughter, Carrie, and Gary Spicer, testified about how they looked into the vacant, weed-covered field and saw "muzzle flashes" and a man crouched with a rifle, wearing a baseball cap and a dark leather jacket.*
>
> *Mr. Spicer told how the sniper tossed the rifle in the trunk of his car and drove away from the field while the gunshots still echoed in the neighborhood.*

Robert L. Van Sciver, one of the most highly respected defense lawyers in the state, led Franklin's defense team. The Franklin defense claimed that his visual disability would have prevented him from the kind of long-range kills he was accused of. They supported this assertion with testimony from an ophthalmologist and a military sniper. Van Sciver said that witnesses like Spicer said the man they saw was not wearing glasses. When I read this, I wondered whether it mattered, given the enlarging power of a telescopic gunsight.

Both of Franklin's sisters came to Salt Lake City for the trial and Marilyn Garzan stayed through to the end. She told a UPI reporter her brother had developed a deep hatred of Blacks because of harassment and assaults family members suffered growing up in their predominantly Black neighborhood. This didn't square with what we had found while researching his background. "The sister," the article stated, "said she had learned to consider Blacks as individuals, not as a collective group." Apparently Franklin had learned no such thing.

Despite Van Sciver's desire to put his client on the witness stand, Franklin decided not to testify in his own defense, which was his absolute right.

When it came time for the summations, though, Franklin did offer a few off-the-cuff comments. When Snarr mentioned Herrera's, Hawley's, and Cooper's testimony during his summation, Franklin shouted out, "They're liars, all liars! Why don't you make them take a polygraph test? They're manipulated by the FBI!" and declared, "I don't feel like going to a firing squad because of some liars!"

Judge Jenkins ordered him removed to a small holding cell, where a speaker was rigged up for him to hear the rest of the trial. As security guards led him out past the judge's bench he screamed, "I won't be quiet when there's a bunch of liars saying those things!"

Van Sciver, pacing before the jury with his hands plunged deep into his suit jacket pockets, attempted to rebut each of Snarr's points, noting that the medical testimony indicated eight or nine bullet wounds, while the police found only six empty shells. Did that mean there was more than one gunman at a different location? He said that several witnesses thought they heard shots coming from different directions. Taken together with the apparent angle of the shots, he declared, "As a citizen of this country, I am offended that the government in this case has failed to explain away the physical evidence. The government's evidence simply did not fit together."

He suggested the investigators were totally swayed by witness interviews and accused them and the prosecutors of "tunnel vision": "From the totality of the ugliness of the offense of August 20, 1980, we've got to convict, we've got to punish somebody . . . and they rely on the people who share Joseph Paul Franklin's secrets."

Following closing arguments and Jenkins's instructions, the all-white jury of ten women and two men began deliberations at 4:15 on Tuesday, March 3. They deliberated for thirteen and a half hours before returning the next afternoon with a verdict of guilty on both counts of having violated the civil rights of the two victims. In contrast to his earlier outbursts, Franklin appeared impassive as the court clerk read the verdicts. Jenkins set sentencing for March 23.

Ted Fields's father, Reverend Theodore Fields, told reporters and supporters, "I feel justice has been done, and I feel happy about it."

Richard Roberts, an African American attorney with the Department of Justice's Civil Rights Division who had assisted Snarr with the prosecution, called the case a signal that his department would "vigorously prosecute any civil rights cases in the country."

In tears, Marilyn Garzan left the courtroom saying, "It's not over. There's no way it's over."

And as he left the courtroom in handcuffs, Franklin proclaimed to reporters, "I didn't do it. It's a government frame-up. That's what I've said all along."

The Associated Press reported, "Franklin, who has maintained he didn't kill anyone, has told reporters earlier he felt the Black youths deserved to die for 'race mixing.'"

Van Sciver said he would appeal the verdict. "I was disappointed," he admitted. "We tried very hard and I think there were problems with the state's evidence." At about the same time, Deputy County Attorney Robert Stott said his office would now proceed with state murder charges against Franklin.

At sentencing, UPI reported that as Snarr commented about all the lives that had been shattered by the crime, Franklin, standing in the center of the courtroom with Van Sciver, dressed in a blue pin-striped suit and striped tie, his hair neatly trimmed and parted in the middle, yelled, "Got any more lies about me, you little faggot—you and that trained ape you've got lying for you," referring to Roberts. As he said this, he leapt at the prosecution table and dove at Roberts, knocking a glass and pitcher of water over, splashing the two prosecutors and dumping ice into the lap of FBI agent Curt Jensen. Ten U.S. marshals and Salt Lake County sheriff's deputies wrestled Franklin to the carpet in front of the judge's bench. It took several minutes to subdue him.

"Put him in irons," Jenkins directed.

Marshals cuffed and shackled Franklin and hauled him upright before the bench. He complained that the handcuffs were too tight and were cutting off the circulation in his hands. The marshals loosened them, but every time Jenkins started to speak, Franklin complained again.

"You'll just have to live with it, Mr. Franklin," Jenkins said. "I want your full attention. This whole thing has been a tragedy from beginning to end."

Judge Jenkins noted Franklin's unhappy youth, saying, "I suppose that's an explanation, but not an excuse for what happened here." He said it was still not too late to change his life in prison.

"Not if I'm sentenced for something I didn't do," Franklin yelled back. "This whole thing is a farce!"

The judge then handed down two life terms, the maximum for each federal charge, which in practical terms meant anywhere between ten and sixty years in federal prison.

Franklin responded, "You are nothing but an agent of the Communist government, you bastard!" as marshals hauled him upright before the judge's bench.

After a three-year reign of terror, Joseph Paul Franklin was behind bars for what would turn out to be the rest of his life. But his story was far from over. And his horrific legacy was just beginning.

PART 2

INTO THE MIND OF A MONSTER

CHAPTER 11

Driven by the results from the Franklin case and our work in other areas, the profiling program started taking off around the same time that Franklin was heading off to prison.

Headquarters took notice of the accuracy of my fugitive assessment and the advice I gave on how to interact with Franklin on the flight to Utah, certainly boosted by Robert Dwyer's complimentary report. And it was gratifying that Dave Kohl's faith in the Behavioral Science Unit was validated. Jim McKenzie, the FBI's assistant director for the Training Division, which made him the head honcho at Quantico, had always been a supporter, and he took great pride in our accomplishments. Larry Monroe, who had been one of the outstanding instructors at the academy, was now the BSU chief, and he had come to realize that profiling and investigative consultation could be a significant and useful addition to our services.

But the case that really put us on the map was in many ways almost a "photo negative" of Franklin's.

By the time my unit was called into the Atlanta child murders case, the Georgia capital was a city under siege. Public Safety Commissioner Lee Brown had set up a Missing and Murdered Task Force with more than fifty members, but the murder and disappearance of African American children continued. In September 1980, Mayor Maynard Jackson asked the White House for help, and on November 6, Attorney General Benjamin Civiletti ordered the FBI to investigate whether any of the missing children had been kidnapped and moved across state lines, thereby triggering federal law and jurisdiction.

But there was another aspect to the case that made FBI involvement less of a stretch. With sixteen apparently linked killings in one of the most progressive cities in the South, this looked to be a Classification 44: violation of civil rights. It was designated FBI Major Case 30, codenamed ATKID. Was this a conspiracy to commit genocide on the Black population, as Franklin had hoped to inspire? Was it the Ku Klux Klan, the American Nazi Party, or some other hate group moving from the blustering talk Franklin scorned to the action he longed for? If so, then Atlanta and maybe other southern cities could be tinderboxes ready to burst into conflagration.

Roy Hazelwood and I flew down to Atlanta in January 1981. Roy was the logical choice to join me, a brilliant special agent and academy instructor who was the Bureau's foremost expert on interpersonal violence. Of all the academy instructors, he was doing the most profiling, taking on many of the rape cases that came to the unit. When we got to Atlanta PD headquarters, we spent considerable time going through the extensive case files—crime scene photos, victimology, descriptions of what each child was wearing when found, statements from witnesses in the area, autopsy protocols, etc. We talked to family members of the missing and murdered children and had the police drive us around each of the neighborhoods where the children had disappeared, and to each of the body dump sites.

The first thing we determined was that these were not Klan-type crimes. If you study what we call group cause hate crimes, going all the way back to post–Civil War days, they tend to be highly visible, highly symbolic acts, designed to instill terror. If a hate group had been responsible, it wouldn't have taken investigators months to start linking cases.

The second thing we noted was that the body dump sites and the areas where the victims were last seen were in predominantly or exclusively Black areas of the city. A white individual, much less a white group, could not have prowled these neighborhoods without being noticed. The police had canvassed and interviewed extensively and had no reports about white people in unusual places. Many of these areas had street activity around the clock, especially in the warm-weather months, so even under cover of night a white man could not have gone unnoticed on several occasions. Unlike Franklin's crimes, these were not committed from a distance. The victims had been abducted and their bodies had been moved after death. We concluded that the killer was a single African American male, and based on the age of the victims, we thought he'd be in his early to mid-twenties. He'd have some kind of hustle to lure these kids from poor neighborhoods, maybe claiming to be an athletic or music scout, or even a cop. He'd be a police buff to compensate for his own inadequacy, and probably drive a police-type vehicle and have a police-type dog. Based on the victimology, he would probably be homosexual and might also resent being Black himself.

The differences between Franklin's crimes and the Atlanta killer's gave us important insight. Being largely mission driven, Franklin wanted no personal contact with his victims; he certainly didn't want to meet or personalize them. To him, they were merely ciphers for their race or religion. Once he committed each crime, he wanted to escape and get out of town as quickly as possible. The Atlanta UNSUB was emotionally invested in his relationship with his victims, how they perceived him, and his power over them.

None of our observations about the killer made us very popular with the public, the media, or the Atlanta PD, most of whom still thought this had to be a series of hate crimes. It also didn't make us popular when we declared that while many of the murders were linked, we didn't think they all were. The two female victims didn't seem to be related to this killer, or to each other. There was even evidence in a couple of the cases that they might have been family murders.

It took months before there was any resolution. During that time, several incidents made us realize that the UNSUB was following and reacting to the media coverage, which allowed us to manipulate him. We had the medical examiner announce that they were finding fibers and other evidence on the bodies, which was true. As we anticipated, this prompted the killer to start dumping bodies in the local rivers to eliminate that evidence. On May 22, 1981, the last day of a surveillance experiment on all the local rivers, Wayne Bertram Williams, a twenty-two-year-old African American, self-styled music producer, was observed dumping a body off the

Jackson Parkway Bridge over the Chattahoochee River. He fit the profile in every key respect, including owning a German shepherd and a police-type car with a police scanner, and a previous arrest for impersonating a law officer. By June 21, the evidence was compelling enough to arrest him.

The story made national headlines, as did the role behavioral science had in the arrest. During Williams's subsequent trial, the prosecutor, Fulton County assistant district attorney Jack Mallard, asked for our help in strategizing ways to get Williams to show his true character to the jurors. I thought he might be arrogant and self-confident enough to take the stand, and that was where our opening would be.

As I'd predicted, Williams did testify in his own defense. After several hours of cross-examination, Mallard had caught him up in a number of glaring inconsistencies, but Williams maintained the same calm, mild-mannered attitude that he'd shown to the public since his arrest. Then, as we had rehearsed, after leading up to one of the murders, Mallard moved in close to the witness box, put his hand on Williams's arm, and in his low, methodical south-Georgia drawl, asked, "What was it like, Wayne? What was it like when you wrapped your fingers around the victim's throat? Did you panic, Wayne? Did you panic?"

In a low, weak voice, Williams said, "No." Then, realizing what he had done, he flew into a rage. He pointed his finger at me and screamed, "You're trying your best to make me fit that FBI profile, and I'm not going to help you do it!"

That was the turning point in the trial. And it was also a turning point for the FBI profiling program. The success in apprehending Franklin, and then the publicity surrounding

Atlanta, made believers out of the headquarters brass. We had proved our effectiveness in helping catch criminals and put them behind bars.

A number of notable cases followed. We flew up to Anchorage, Alaska, and assembled a profile that convinced a judge to issue a search warrant on Robert Hansen, a baker in his mid-forties whom the police suspected of the murders of prostitutes found shot to death in remote wooded areas. We described a proficient hunter who had tired of shooting animals and was now imitating "The Most Dangerous Game," flying women he had picked up out into the wild and hunting them for sport to compensate for the way he had been treated by women in the past.

In June 1982, I was asked to provide a profile and analysis of the 1978 murder of twenty-three-year-old Karla Brown of Wood River, Illinois, a beautiful, all-American girl who was found stripped, her hands tied with an electrical cord, and her head shoved into a ten-gallon barrel filled with water in the basement of the house she was about to move into with her fiancé, Mark Fair. Coming up with a visualization of the crime based on all the evidence, I profiled an UNSUB who matched two of the individuals police had questioned. Karla's body was exhumed, and bite marks matched one of the suspects, adding to the evidence against him. He was convicted and sentenced to seventy-five years in prison.

These were only two of the hundreds of cases we were now handling each year. Jim McKenzie, one of our greatest advocates, sold headquarters on the need for "more John Douglases" to handle the ever-growing caseload, even though it meant stealing slots from other programs. That was how I got my first four full-time profilers: Bill Hagmaier, Jim Horn, Blaine McIlwaine, and Ron Walker. Shortly after that, Jim Wright and Jud Ray came onboard. That group represents the foundation of what came to be the Investigative Support Unit (ISU).

And we continued our study of incarcerated serial killers and violent offenders. Around this time, I was collaborating on an article for the *FBI Law Enforcement Bulletin* about lust murder with Roy Hazelwood. Roy introduced Bob Ressler and me to Dr. Ann Burgess, a professor of psychiatric nursing at the University of Pennsylvania who was also associate director of nursing research for the Boston Department of Health and Hospitals. Ann was widely known as an authority on rape and its psychological consequences, and she and Roy had done some research together. Ann was impressed by the serial killer study Bob and I had embarked on, and we agreed to work together. In 1982, she secured a four-hundred-thousand-dollar grant from the government-sponsored National Institute of Justice to formalize our study, and, with our input, developed a fifty-seven-page assessment protocol that we would fill out after each of our subsequent interviews. That study would further our understanding of the violent criminal mind and support the type of behavioral profiling and criminal investigative analysis the FBI practices to this day. It led to the publication of the book *Sexual Homicide: Patterns and Motives* that the three of us coauthored and eventually to the *Crime Classification Manual* (*CCM*), which we intended to be the equivalent for law enforcement personnel of the American Psychiatric Association's *Diagnostic and Statistical Manual of Mental Disorders* (*DSM*). It presents the range of violent crimes and motives and helps investigators identify the type of crime and why the offender might have committed it. Some, like bank robbery, are easy to understand. Others, like torture murder, can be far more perplexing. What we ultimately wanted to do was translate our psychological and behavioral research into concepts and terminology that would be useful to law enforcement. For example, telling a detective that an UNSUB appears to be a paranoid with schizoid tendencies may not help him or her very much in an investigation. Saying the offender is organized, however, could, with other profile descriptors, help them narrow down a suspect list. *CCM* has been revised several times and is now in its third edition.

And yet, as I moved on from Joseph Paul Franklin, as I interviewed other killers and my unit gained notoriety within the bureau, there was something about Franklin and his crimes that continued to haunt me. The more killers I'd talked to, the more I'd come to understand their psyches and root motivations, the clearer it became that Franklin's personality represented something deeper and even more disturbing. Every serial killer and violent predator is trying to compensate for their own inadequacies and to vent their anger and resentment at the world, or the part of it that they perceive has denied them their due. But as I'd suspected when Dave Kohl first asked me to study his case and now felt even more strongly, Franklin, even behind bars, remained far more dangerous than most. His unwavering dedication to fomenting hate made him a potential inspiration and symbol to others with similar orientation. Infamous killers can inspire occasional sick minds like theirs, but it stops there. Joseph Paul Franklin had the capacity to inspire untold numbers of like-minded young men to embark on the same journey he had: from words to action, from resentment to murder. I knew that one way or another, I was not finished with Joseph Paul Franklin.

I wasn't the only one interested in him and his motivations. By the late 1980s, Kenneth Baker, a distinguished Secret Service agent who had served on presidential details and who, like me, had a doctor of education degree based on studies and methodologies he had developed in the field, temporarily came over to my unit for a joint Secret Service–FBI project on assassins. One of the first interviews he conducted for this joint project was with Mark David Chapman, imprisoned at the time at the Attica Correctional Facility near Buffalo, New York, for the December 8, 1980, murder of ex-Beatle John Lennon. Chapman, then twenty-five, had shot the forty-year-old Lennon at close range as he and his wife, Yoko Ono, were returning from a recording session to their apartment in the Dakota, on Seventy-second Street at Central Park West in Manhattan. He fired five hollow-point bullets from a Charter Arms .38 Special revolver, four of which hit the rock and roll superstar.

We uncovered a range of critical similarities and differences between an up-close-and-personal assassin, like Chapman or would-be assassin John Hinckley, who tried to kill President Ronald Reagan outside the Washington Hilton hotel less than four months after the Lennon assassination, and a sniper-style assassin like Franklin. My unit was heavily involved with the Hinckley investigation.

Chapman and Hinckley, like Franklin and virtually all assassins we have studied, felt deeply inadequate. Eventually, that inadequacy became overwhelming and had to be acted upon. Hinckley got it through his shallow and misguided head that he could impress the object of his romantic fantasies, actress Jodie Foster, by killing the president and having her join him on a commandeered airplane, from which they would fly off to some undefined land of happily-ever-after. Both Chapman and Hinckley looked to the book, *The Catcher in the Rye*, J. D. Salinger's saga of disillusioned youth, for their personas, just as Franklin looked to *Mein Kampf* for his. What every assassin convinces himself—from Brutus and Cassius, through John Wilkes Booth, Lee Harvey Oswald, and up to the present—is that things will get better as a result of his bold and historic act.

When Ken Baker interviewed Chapman at Attica, what he found was someone who had a strong emotional connection with his target on a superficial level. He collected all of the Beatles' and Lennon's records and even went through a string of Asian girlfriends in imitation of Lennon's marriage to Ono. Ultimately, Ken learned from the interview, Lennon became an impossible model for Chapman to live up to, so he fabricated in his own mind a reason to kill him: the contrast between Lennon's preaching of love and peace and eschewing material possessions with his glamorous and expensive lifestyle, and Lennon's supposed religious sacrilege after Chapman's born-again embrace of Christianity.

But these were mere excuses, like Franklin's having other groups to blame for his own shortcomings. Chapman could no longer deal with the disparity between himself and his erstwhile hero, so he had to kill him. This would achieve a twofold aim. Lennon would no longer be around for him to have to compare himself with, and at the same time, his name would be forever linked with Lennon's. He did not have the power to become like John Lennon, but he did have the power to destroy him. Whatever he had failed to accomplish in his miserable life thus far, he was now no longer a nobody.

We pursued the assassin study with the same rigor we had devoted to serial killers and sexual predators, and it yielded important insights into their psyches. You might think someone like Arthur Bremer—the man who shot and paralyzed former Alabama governor George Wallace while he was campaigning for the presidency in a Laurel, Maryland, shopping center in 1972 and whom I interviewed for the project—might be the diametric opposite of Joseph Paul Franklin. After all, he tried to kill one of the most powerful symbols of segregation and racial bigotry, the man who stood in front of Foster Auditorium at the University of Alabama to prevent African American students from registering.

But the more I learned, the more I came to realize that Bremer was simply trying to prove his own worth. He had previously stalked President Nixon for several weeks but could never get close enough. So, in desperation, he trained his sights on a more accessible target. When I interviewed him, I found he had no particular issue with Governor Wallace at all.

The one factor that is clearly different between Franklin and these others is that Franklin was not up-close-and-personal, which signifies that he intended to get away with his crimes rather than go down in flames as a martyr or engage in some ridiculous fantasy like Hinckley. But as we will see, that doesn't mean he figured everything out ahead of time.

The one ironclad procedure we'd developed for all of our prison interviews was that before going in, we had to learn everything we possibly could about the criminal and his crimes. And we had seldom encountered one with as much proficiency, mobility, and determination as Franklin. His background was complicated, and his list of crimes was long and probably incomplete. But as I started considering Franklin as a subject for our study and began reading about his life behind bars and all that had transpired since he was sent to prison in 1981, I realized that our understanding of him as a killer was far more incomplete than I ever could have realized.

As it turned out, there was a lot more to uncover, learn, and study about his life and crimes before we attempted to meet him face-to-face.

CHAPTER 12

In the immediate aftermath of Franklin's conviction in 1981, law enforcement authorities around the country with pending cases against him had to make decisions about which of the multitude of cases against him would come next.

Back in Utah, Franklin's defense attorney Van Sciver had filed an appeal based, along with some other procedural issues, on the fact that Judge Jenkins allowed evidence that Franklin had Maced a mixed-race couple in 1976. Part of the defense attorney's legal reasoning was that the incident had been too far in the past. But the other part was more interesting and, I thought, turned normal defense strategy on its head. Van Sciver claimed that the Mace attack should not have been introduced to show motive, because it was well established that Franklin was a racist; he was merely denying his participation in the murders in Liberty Park, and therefore the Mace incident was prejudicial to him receiving a fair trial strictly relating to the charge at issue.

While Franklin was in the Medical Center for Federal

Prisoners in Springfield, Missouri, for evaluation and assignment into the federal corrections system, there was jockeying among the various jurisdictions in which both state and federal charges were pending as to who would try him next and who would give up a place in the queue. When a suspect is charged with murder in more than one jurisdiction, it is not uncommon for there to be bargaining and maneuvering to decide which one gets first crack at him. The Justice Department decided to drop federal bank robbery charges in Arkansas and Kentucky so that Franklin could be turned over to state authorities for prosecution on the murder charges. Utah would be up first.

By the time the state of Utah was ready to try Franklin for the Liberty Park murders, he was on his fourth set of court-appointed defense attorneys. As if that volatility weren't enough, Franklin also requested that the presiding judge, Third District Court judge Jay Banks, disqualify himself from presiding because he "has demonstrated a prejudiced and bigoted attitude toward the defendant . . . and is a complete and total lackey and stooge of the federal government." Not only is this not the greatest way to endear oneself to a judge, I thought it was rather ironic that this self-proclaimed racist would comment on someone else's supposed bigotry. In the same hearing, Franklin asked that four newspaper and television reporters be barred from his trial because they had slandered him and were "paid FBI informants and snitches." Franklin's inherent paranoia was on clear display, but at the same time, he had become a hero to white supremacist groups, who were making death threats against the prosecution team.

In June 1981, Judge Banks ruled that trying Franklin on

murder charges in state court after he had already been tried in federal court for civil rights violations against the same two victims did not constitute double jeopardy. After first turning down Franklin's request to serve as his own attorney and reminding him of the old adage "He who represents himself has a fool for a client," Banks reversed himself the next week and allowed it. The condition was that he conduct himself in a proper manner in the courtroom and accept assistance from actual legal defenders. Former Utah attorney general Phil Hansen agreed to take the case on Franklin's behalf, as did David E. Yocom and D. Frank Wilkins, a former justice of the Utah Supreme Court.

The trial began on Monday, August 31, with jury selection. The final panel was made up of seven men and five women, all white. Opening statements were delivered on Thursday, September 3. Franklin, attired in a dark blue three-piece suit and striped tie, his longish, reddish-blond hair neatly combed, claimed he had just been "in the wrong place at the wrong time," passing through Salt Lake City on his way to San Francisco.

Micky McHenry, now going by the name Micky Farman-Ara, the prostitute Franklin had picked up shortly before the murders, testified in the trial. Terry Elrod and Karma Ingersoll told about the shootings while they were jogging with Fields and Martin. When Franklin's ex-wife Anita Cooper took the witness stand and said that her former husband had confessed to the Liberty Park killings, he cross-examined her, saying, "You made that up to get back at me, didn't you? Isn't that a malicious lie, a cold-blooded lie?"

After a fifteen-day trial in which seventy-five witnesses

took the stand, the jury received the case on Friday, September 18, and deliberated for about six and a half hours before returning with a guilty verdict. Franklin's sisters, Marilyn Garzan and Carolyn Luster, were in the courtroom and sobbed quietly when they heard the verdict. Franklin himself showed no emotion.

During a recess in the sentencing phase the following Wednesday, Franklin once again attempted an escape, this time from a holding area, by sneaking into the key-operated elevator that linked the Metropolitan Hall of Justice to the county jail through a basement tunnel. He used a screwdriver he'd managed to secure, probably from another inmate, pried open the control panel outside the elevator door, and hot-wired the mechanism with a dime and a paper clip. It always interests me when I see criminals like Joseph Paul Franklin, who had singular lack of success or interest in holding down regular jobs, being so resourceful in robbing banks, killing with a sniper's precision, and even being able to hot-wire a car or an elevator.

Police and marshals fanned out in the building and on the surrounding streets, which were cordoned off with a frantic hunt in progress. Franklin, meanwhile, took the elevator down two floors to another courthouse holding area. John Merrick, the jail guard who had sat behind Franklin during the trial, found and confronted him as he was attempting to remove the pins from the hinges on a door leading to the public hallway. Guards also noticed he had pulled out a small grating in the elevator, apparently considering escaping through the elevator shaft like in the movies. After they cleaned him up, Merrick and an angry Sheriff Pete Hayward brought him back to the courtroom, without the jury knowing what had happened. He had enjoyed all of fifteen minutes of freedom.

Because Utah had the death penalty, the jury also had to decide whether Franklin deserved it. In pleading against the death penalty, defense attorney Yocom pointed out that Franklin continued to maintain his innocence, implying that they would be taking a big chance in opting for capital punishment, which he characterized, correctly, as an "irretrievable, final act." Then he said, "Joseph Paul Franklin is an intelligent, religious, humorous, useful human being. And given the chance by you, I'm sure he could make a contribution to the world, this country, this society. The defendant has enriched my life. I'm sure he could do it for another."

Huh? Intelligent? Religious? Humorous? Useful? Are we talking about the same Joseph Paul Franklin? I guess he was intelligent in that he was able to kill people and rob banks without getting caught for several years. Perhaps you could say he was religious in that he thought Jesus Christ supported his deadly mission. If you call laughing at the killing of African Americans humorous, I suppose you could check that one off. And if you wished to see Adolf Hitler's vision finally carried out in 1980s America, he was useful. But that is not how I would ever construe those terms. Oh, and just for the record, so far as I know, Franklin never did enrich anyone else's life, though he sure ended enough of them.

Stott countered by saying that the only ones who deserved any sympathy were the two victims, that sniper killing in this manner was a particularly cowardly act. "The joggers, the young victims, were joking, they were laughing, they were kibitzing, they were just enjoying their lives. They didn't know what hit them, they didn't know who hit them, and they didn't even know why they were hit. From the moment Joseph Paul Franklin stepped into that field and hid behind that mound of dirt with a rifle, he had made the decision to subject himself to the death penalty. Nobody forced him to make that decision."

He went on to declare, "The true victims of circumstance in this crime were Dave Martin and Ted Fields; they were the ones who happened to be in the wrong place at the wrong time."

After two hours of deliberation, the jury returned to the courtroom, split eight to four in favor of the death sentence. Since the verdict would have had to be unanimous, Judge Banks imposed a life sentence for each of the two murders.

In imposing the sentence, Banks stated he would recommend that Franklin never be paroled. Franklin began cursing the judge, and as he was being led out of the courtroom by the marshals, he shouted, "You're the one who ain't got no morals!"

After the sentence was announced, reporters asked Marilyn Garzan her feelings about her brother's escape attempt. "It's a shame he didn't make it," she replied.

A few days later, back home in Montgomery, Alabama, Carolyn Luster told the *Birmingham Post-Herald* that Franklin had planned the escape two days before. "It was not a spur-of-the-moment thing. If he had got out, they never would have found him."

Under the terms of the writ by which Franklin was remanded to the state for the trial, since he was not sentenced more harshly, i.e., to execution, he was returned to the Medical Center for Federal Prisoners in Missouri and federal custody to serve out his civil rights sentence.

"The federal authorities have better facilities to deal with Joseph Paul Franklin," Utah prosecutor Stott said.

ON JANUARY 31, 1982, FRANKLIN WAS TRANSFERRED FROM THE MEDICAL Center in Springfield to the United States Penitentiary at Marion, in southern Illinois, which had been opened in 1963 to replace the decrepit and expensive-to-maintain Alcatraz maximum-security prison in San Francisco Bay. By the time Franklin arrived, Marion had the reputation of housing the most violent and unmanageable inmates in the federal corrections system. Prisoners were only allowed out of their cells about an hour and a half a day, and outside recreation was even more severely limited.

Franklin was a fairly well-known con by the time he arrived, and given the prison's large African American population, his reputation proceeded him. Just three days after he came to Marion, Franklin had returned from dinner and was hanging out briefly in another prisoner's cell. A group of Black inmates surrounded and cornered him, then stabbed him fifteen times in the neck and abdomen with an ice-pick-like weapon made from a can. There were several guards nearby, but they said they did not see the attack. Franklin was rushed to Marion Memorial Hospital and taken into surgery.

"We really don't know if it was racially motivated or not," FBI special agent Robert Davenport of the Springfield, Illinois, field office stated. Though he was trying to be careful, I, on the other hand, didn't see any other possibility. On March 3, Davenport said his office had given the results of its investigation to the U.S. attorney in East St. Louis, Illinois, but "prosecution was declined because Franklin could not identify his attackers and there were no witnesses who could identify the assailants."

Franklin was returned to Marion and placed in a special basement unit that was kept apart from the prison's general population. It was called the K Unit. It had somewhat larger cells with their own toilets and showers, in which the prisoners spent twenty-two hours each day. For security, no two men on the unit were allowed out of their cells at the same time.

While Franklin's trail of bloodshed had stretched across multiple states, with the murder case in Utah officially closed, the case that became highest priority from Washington was his potential involvement in the shooting of Vernon Jordan. FBI director William Webster told the *Los Angeles Times* that Jordan had been gunned down by one or more people who had been stalking him. He said it was a calculated act and that the Bureau had discounted any nonpolitical or personal motives for the ambush shooting.

Franklin remained a principal suspect. The FBI had determined through matching handwriting on motel registration cards that Franklin was in the Fort Wayne area at the time of the crime. Through his lawyer, Franklin refused to submit to a polygraph. I have never put much faith in these so-called lie detector tests, and in most jurisdictions, they are not legal evidence. I was convinced that Franklin made a regular habit of lying through his teeth, so as with other sociopaths I have hunted, I didn't think it would be any more challenging for him to lie to a metal box than to lie to other human beings. We've long established the idea that these kinds of people react differently, so I didn't see that a polygraph would prove anything one way or another.

For his part, Vernon Jordan seemed to gather new strength and determination when confronted by darkness and violence, much as did his inspirational ideal, Martin Luther King Jr. Following a three-month recovery, during which he lost forty pounds, Jordan told a news conference at the National Urban League's New York headquarters, "It is an indescribable experience. All of a sudden there you are stretched out on the pavement bleeding and with some notion that it might be curtains." Declining to speculate on who shot him, Jordan said, "I have not spent the last few months concentrating on who did it or why they did it. I was just trying to get through the next operation or the next shot, the next bad-tasting medicine, the next exercise some nurse demanded and which I did. . . . Violence is nothing new to Black people. We have been its victims from the time the first slave ship landed on these shores. We recognize we are vulnerable in a society in which racism still thrives."

But the legacy of the crime loomed large in other ways. According to an account by the Associated Press reporter Masha Hamilton, Fort Wayne was still agonized over the Jordan shooting. "Jordan's health apparently has returned," she wrote on May 29, 1981, "but this northeastern Indiana city of 175,000 won't recover fully until a suspect is arrested, city officials and Black leaders say." She went on to describe "a sense of paranoia that exists in the local Black community," underscoring how one malign individual can transform an entire region.

Franklin's attorney David Yocom, though, considered the government's pursuit of the Jordan shooting "ridiculous." When a grand jury of the U.S. District Court in South Bend, Indiana, returned an indictment on June 2, 1982, against Franklin for violating Jordan's civil rights—shooting him because of his race and color—Yocom called it "a waste of taxpayers' money."

"It's ridiculous that a man serving four consecutive life sentences . . . would even be considered for being prosecuted for a maximum ten-year [sentence]," he told the AP. "The Justice Department must really be looking for points with the Black people." Residents of Fort Wayne and the surrounding area, on the other hand, were relieved and gratified that the crime and alleged perpetrator were finally facing justice, according to newspaper reports.

On June 11, Franklin pled not guilty to the charge and Judge Allen Sharp set a trial date in August.

While newspapers around the nation, from the *New York Times* to the Lafayette, Indiana, *Journal and Courier*, regularly referred to Franklin as "an avowed racist" and "a drifter," in a pretrial hearing Judge Sharp warned prosecutors they would not be allowed to rely on any of Franklin's other proven or alleged crimes in presenting their case. "There will be no retrial of the Utah case in my courtroom," he declared. Sharp was apparently skeptical of the strategy to go forward with a federal civil rights action because there didn't seem to be enough solid evidence to pursue a state murder charge. Toward the end of the trial he was quoted by UPI as saying the case was "pushing federal court jurisdiction close to its constitutional limits." And he was concerned enough about the public passion surrounding the case that he ordered all spectators to be physically screened and pass through metal detectors.

Admittedly, going the civil rights route on a case like the Vernon Jordan shooting meant going through some legal contortions, which seemed to me even more challenging than a straight murder trial. The prosecution would have to prove not only that Franklin was the one who pulled the trigger, but that he did so to *prevent* the victim's use and enjoyment of public facilities *because* of his race.

The prosecution was headed by Barry Kowalski, an antiwar activist who nevertheless enlisted in the Marine Corps and led a platoon in Vietnam because he didn't think only poor people should have to go and fight. He was already securing his near legendary status as a civil rights attorney and "the Justice Department's pit bull." He would go on to secure convictions against KKK members James Knowles and Henry Hays for the 1981 abduction and lynching of nineteen-year-old Michael Donald; the 1988 conviction of neo-Nazis for gunning down Jewish radio host Alan Berg outside his Denver home in 1984; and the 1993 federal conviction of two white LAPD officers for the beating of Rodney King in 1991 after they had been acquitted in local court.

Still, Kowalski encountered roadblocks from the start. The judge denied his motion to introduce evidence from the Salt Lake murders and the time Franklin reportedly sprayed Mace at the mixed-race couple in Maryland in 1976. Kowalski conceded to the all-white, eight-man, four-woman jury at the outset that there was no eyewitness to the crime and that the murder weapon had never been found, but asserted that there was a mountain of circumstantial evidence and that two prison inmates would testify that Franklin had bragged to them about shooting Vernon Jordan. One of them would say that while watching television coverage of the Atlanta child murders, Franklin expressed approval of killing the Black children. Franklin was "so thoroughly obsessed with a passionate hatred," Kowalski declared, that "he attempted to take another man's life."

Specifically, he hammered, "Franklin shot Vernon E. Jordan because Mr. Jordan is Black and was in the accompaniment of a white woman and because he was using facilities of the Marriott Inn."

On the second day of the trial, Jordan himself came in to testify, but later conceded he didn't have much to contribute in his testimony other than to describe the sensation of "sailing through the air and a rather dramatic pain in my back" as the bullet struck him and wondering whether he was dreaming. He said he didn't even hear the gunshot sound. He repeated that he had no idea who had attacked him or why. "It seemed an eternity as I lay there. It seemed like help would never get there," he said.

His surgeon, Dr. Jeffrey Towles, testified, "I think he was just about as close to dying as one can come without dying," describing the wound as large enough for him to place his fist in, and how close the bullet came to Jordan's spinal cord.

Walter White, a supermarket security guard, told the jury he overheard Franklin asking a clerk if President Jimmy Carter was coming to Fort Wayne, and about Jordan's medical condition. White testified that Franklin told the clerk the shot that hit Jordan "was about perfect. If it had been a little bit different, it would have gotten him just right."

Mary Howell, a maid at a Fort Wayne motel, identified Franklin as the man she knew as "Joe," who, a few days before the Jordan shooting, "said he couldn't understand why the manager would rent rooms to so many Black people," and that "all the maids should carry a gun to protect ourselves against Black people."

Steven Thomma, a former newspaper reporter in Fort

Wayne, testified to a telephone conversation he had had with Franklin in the Salt Lake County Jail. He said Franklin claimed never to have been in Fort Wayne and that the motel registration cards the prosecution claimed placed him in the city at the time of the shooting were forgeries meant to frame him because of his racist views. Franklin did admit to selling a .30-06 rifle after placing a classified ad in the *Cincinnati Enquirer.*

Peggy Lane, an *Enquirer* reporter, said she had come across the ad run on June 7 and 8, 1980. She traced the phone number listed to a motel in Florence, Kentucky, the suburb where Franklin was first arrested and escaped the police station.

Robert Herrera, Franklin's former jailhouse neighbor who'd provoked so much anger from the defendant when he testified at Franklin's federal trial in Salt Lake City—where he'd been found guilty of having violated the civil rights of Ted Fields and David Martin III—testified that Franklin admitted to the shooting in "Fort Something, Indiana." As with the earlier trial, the strength of Herrera's testimony came with something of a "jailhouse snitch" caveat: although he said he wanted nothing in return for his time on the stand except a letter to the parole board, it came out that on another case, he was paid for his testimony.

Another witness to take the stand against Franklin was a white supremacist named Frank Abbott Sweeney. Back in the general population at the Medical Center for Federal Prisoners in Springfield, Franklin and Sweeney befriended each other. Franklin was impressed that Sweeney had once served in the Rhodesian army, which had been one of Franklin's aspirations, where he dreamed of killing Black people at will. According to what Sweeney later reported, Franklin told him of his exploits traveling around the country targeting African Americans. He said he would sometimes venture into Black neighborhoods to scout them out wearing an Afro wig and blackface. Knowing Franklin's predilection for committing his crimes at a distance, I suspected he stayed in his car on such occasions and never got close enough for anyone to see how ridiculous his disguise must have looked. When Sweeney was paroled, he contacted the Cincinnati Police Department. Investigators were wary of him and his racial views but were apparently impressed that he felt killing the two teens in Cincinnati went too far.

On the stand, Sweeney related how Franklin had told him about shooting someone important in Indiana.

"Vernon Jordan?"

"Yes," Sweeney said Franklin confirmed. "'I shot him, but he wouldn't die. I'm sorry that I didn't shoot the white slut first.'" He further stated Franklin had told him no one had seen him shoot Jordan and that they couldn't tie him to the murder weapon because he "got rid of the piece."

Under cross-examination, Sweeney admitted that prosecutors promised to write a favorable letter for him to his parole board in Newark, New Jersey, but denied he had been paid anything for his testimony. He said he didn't need the money because he had recently inherited $250,000.

Franklin's defense attorney scored a point when he got one of the government's final witnesses, Lawrence Hollingsworth, to admit that his memory of Franklin confessing to shooting Jordan while he and Franklin were both in the Salt Lake County Jail had been "improved," or, as Franklin's court-appointed attorney J. Frank Kimbrough put it, "tainted," by hypnosis. Hollingsworth had related that while both men were watching a television documentary about the 1968 assassination of Dr. Martin Luther King, Franklin bragged that "he shot or killed someone named Jordan."

The fact that Hollingworth was serving time for arson and jury tampering didn't help matters.

With the jury out of the courtroom, Kimbrough moved for a mistrial on the grounds of the "tainted" evidence. Judge Sharp, who had already demonstrated his strict commitment to fairness and objectivity when he admonished the prosecution about retrying previous convictions, turned down the request, stating, "It is not the job of a trial judge to sort through and weigh the evidence. The jurors are the judges of the facts."

That is legally and procedurally true. But testimony from inmates, otherwise known as jailhouse informants or jailhouse snitches, is one of the most problematic areas of trial evidence. Brandon Garrett, a distinguished law professor and author, now at the Duke University School of Law, names it as one of the five common reasons for wrongful convictions—along with false confessions, junk science, ineffective counsel, and bad judging. The issue is a double-edged sword. In many instances, other inmates are the only ones who would get to hear a confession if it was not shared with detectives. But it is undeniable that many informants do so to make a favorable impression on the authorities regarding their own legal situations. Therefore, their testimony is almost always to some degree suspect, and juries can have a difficult time separating fact from self-serving invention. The mere fact that a prisoner may have something to gain from his testimony does not nullify it. But every prosecutor knows that the jury will subject that kind of witness to special scrutiny.

Also outside the jury's presence, Sharp unsealed a motion Franklin had filed asking him to bar Jewish reporters from the courtroom because Jews controlled the mass media and had waged a "campaign of lies and slander" against him. Among his charges was that Jewish journalists had spread "race equality and other communist propaganda." Franklin wasn't a deep thinker, but he certainly was consistent. Not surprisingly, Judge Sharp rejected his motion.

On Friday, August 13, the prosecution rested its case after four days of testimony, and the defense opened its case the following Monday. Franklin took the stand in his own defense, as he had previously said he would do.

"Mr. Franklin, did you shoot Vernon Jordan?" his attorney asked.

"No, I did not," Franklin asserted.

After running through a number of hair-splitting semantics involving his racism, his attorney questioned where he was on May 28 to 29, 1980, if not in Fort Wayne. Franklin responded, "I have no idea."

The only other defense witness was Kenneth Owens, a convicted burglar and prison escapee who said he and Franklin had been close friends at the Medical Center in Springfield and that Frank Sweeney had a bad reputation among fellow inmates for honesty. I took that to mean they didn't place much faith in what he said.

By midday, the defense had rested its case.

In his instructions to the jurors on August 17, Judge Sharp confirmed that they not only had to decide whether Franklin shot Jordan but also that, if they did, they must also decide whether Franklin shot him in order to prevent Jordan from using the public accommodation of the Fort Wayne Marriott Inn, for that was the definition of a civil rights violation.

The jury retired at 12:30 P.M. and deliberated for about eight hours before returning to the courtroom. The court clerk read the verdict: not guilty. Franklin raised his right hand in the V for Victory sign and exclaimed, "All right!"

"Your decision, while a controversial one, in a controversial case, is well within the law and well within the evidence," Sharp said to the jurors as marshals guarded their exit from the courtroom.

Outside the courthouse, prosecutor Daniel Rinzel, who had assisted Kowalski, said, "The jury has considered the evidence and made their decision, and we accept what they did."

When reporters asked him if the federal government would look into other means of prosecuting Franklin, he replied, "This case is finished."

Vernon Jordan had no comment.

Moments after his victory, U.S. marshals surrounded Franklin and escorted him away to continue serving his multiple life sentences at the Marion federal penitentiary. He was back behind Marion's walls within twelve hours.

Later, once Judge Sharp had rescinded his gag order preventing the news media from talking to the jury members, two of them told reporters that most of them believed Franklin actually had shot Jordan, but they couldn't prove it, nor could they prove Franklin's motive of depriving Jordan of his right to use the Marriott's accommodations. One juror said they thought Frank Sweeney's testimony about Franklin's pronouncements in jail was credible. Another said he "didn't believe anything" Robert Herrera said.

As reported by the *Indianapolis News*, "The second juror said deliberations on guilt or innocence came down to looking at the wording in the indictment about the civil rights violations."

"'That was the one point we came down to. If it wasn't for that point, I think he would have been convicted.'"

CHAPTER 13

Within days of the trial and verdict in Fort Wayne, Indianapolis prosecutor Stephen Goldsmith said he saw "very little useful purpose to be served" in proceeding with trying Franklin for the January 1980 murders of Leo Watkins and Lawrence Reese. He said that if he went ahead, he would have to call some of the same witnesses who had just testified in the Jordan case. He was concerned about the believability of the prison inmates. He was also influenced by Fort Wayne prosecutor Arnold H. Duemling, who announced that he did not have enough evidence to indict Franklin for the Jordan shooting in state court. So, as it stood, the Jordan shooting remained open, and the person who had shot him might still be at large.

This is often a difficult decision for a prosecutor. Every murder victim's family I've ever dealt with wants justice for its loved one. And that means charging the alleged offender with the specific crime. It is not enough that he has been convicted of a similar crime against others, as I saw firsthand in the Atlanta child murders when prosecutors felt their best case against Wayne Williams was for two of the cluster of close to thirty murders that took place from 1979 through 1981.

On the other hand, the resources of any prosecutor's office are limited, and if an alleged killer has already been convicted in another court or jurisdiction and has been put away for a long time, the head of the office has to weigh the chances of a conviction against the fallout of an acquittal, whether it is worth his staff's time and energy, and whether a guilty verdict will add any effective prison time to the inmate's already lengthy sentence; in other words, if the end result of a successful prosecution is to impose punishment, are you actually achieving any practical results from another trial? The federal government, encouraged to prosecute by President Jimmy Carter himself, had a different reason for pursuing the Jordan case even though a conviction wouldn't have added appreciably to Franklin's time in prison. What the feds were after, instead, was a highly publicized warning to anyone who considered trying to assassinate another African American civil rights leader.

"When a notorious public crime is committed and the government believes there is sufficient evidence to prosecute the case, we have an obligation to do so," prosecutor Rinzel declared.

There was another issue. For many years, crimes against African Americans, particularly anti-civil-rights-oriented crimes, went largely unprosecuted and unpunished. For a man of Franklin's psyche, it was natural to think that certain crimes were righteous expressions rather than despicable acts. During our study of the assassin personality, we noted that John Wilkes Booth expected to be received as a great hero and savior in the South after he killed Abraham Lincoln. I suspected that something similar was likely going on in Franklin's mind with regard to Jordan.

For Stephen Goldsmith in Indianapolis, if he felt there was any reasonable possibility of Franklin being released from prison while he was still physically able to commit another violent crime, I feel certain he would have wanted to proceed with his case.

Three states to the southwest, though, Oklahoma County district attorney Robert Macy announced his intention to begin extradition proceedings to have Franklin transferred to the federal prison in Oklahoma City so he could stand trial for the 1979 killing of Jesse Taylor and Marion Bresette outside the grocery store by their car, where her three young children waited and witnessed the murder. Since Oklahoma was a death penalty state, Macy figured a conviction might ensure that Franklin never walked the streets again.

He remained in Marion and relatively out of the public eye, with the Oklahoma case never coming to trial. In January 1983, Macy dropped the murder charges, saying the chance of gaining a conviction did not justify the cost of prosecution. Some of the evidence had gone stale and he wasn't convinced all of the relevant police reports were accurate.

Reading about the decision, I have to admit, I was disappointed. I don't believe in the death penalty across the board, but here was one killer who I thought certainly deserved it. We can debate whether capital punishment is a general deterrent; the fact that it is administered so infrequently leads me to believe it isn't much of one. But it is certainly a specific deterrent—no one who has been executed has ever committed another murder. And given Franklin's propensity for escape, I wasn't convinced his life-taking days were over until he'd given up his own.

PROSECUTORS AROUND THE COUNTRY WERE OPTING OUT OF GOING AFTER Franklin, and for many killers that would have been the end of the story—the unsolved crimes they were suspected of committing would remain unsolved. Most convicted criminals consider themselves lucky when other cases against them are unprovable short of a confession or new physical evidence, so they keep their mouths shut. Instead Franklin began doing the opposite.

Though Robert Herrera had associated Franklin with the attempted murder of *Hustler* magazine publisher Larry Flynt, there was no hard evidence he was the sniper who shot Flynt in Lawrenceville, Georgia. Flynt had been in a wheelchair and in constant pain ever since, pain he had tried to quell with massive quantities of narcotics that led to addiction and several surgeries. Eventually, he suffered a stroke that left him with difficulties in speaking.

The case had never gone anywhere because like so many other sniper incidents, no one saw the shooter. Flynt had spent considerable time and money trying to uncover the identity of his attacker. While he was still in the hospital recovering from his two severe abdominal wounds, he told Rudy Maxa, then a reporter for the *Washington Post,* that he was "convinced the attempt on his life was the work of an assassination team with ties to the government. The motive: to silence his inquiry into the JFK killing."

At various times after that he had told interviewers he

thought he had been gunned down by "the same man who shot Vernon Jordan." On another occasion, as reported in the *Atlanta Constitution,* he said "a number of Georgia legislators and politicians were involved in his shooting to stop him 'from exposing what's really going on in this country.'"

In August 1983, Franklin wrote a letter to the Gwinnett County District Attorney's Office, headed by Daniel J. "Danny" Porter. It said, "My name is Joseph Paul Franklin. I shot Larry Flynt. If you bring me to Gwinnett County, I'll tell you about it."

Instead, the next month, Captain Luther Franklin "Mac" McKelvey and Sergeant Mike Cowart went to Marion and spent four hours meeting with Franklin. At first, he said the letter was a hoax, but then agreed to talk seriously with the officers.

"In the initial contact with us, he felt us out, and we felt him out," Cowart told *Constitution* reporter Rob Levin. "He indicated a good bit of knowledge about the case and gave us enough leads that we felt we had probable cause to seek an indictment. The more we checked his background, the more of a suspect he became."

Following the visit, McKelvey had several telephone calls with Franklin. Though he knew Franklin had been reading up on the case in the newspapers, he said the convict provided a piece of information that only the gunman could have known.

As for motive, the original theory seemed likely—that Franklin didn't like *Hustler*'s portrayal of mixed-race couples in its sexually explicit pictorials.

This brings up another question: Why would Franklin suddenly rat himself out if he actually was the shooter?

With his notoriety and the previous brutal attempt on his life, Franklin must have felt that even though he was now in Marion's protective wing, he was a marked man. He clearly stated he wanted to be moved to another, less dangerous prison, possibly in one of the state correctional systems. Being tried and found guilty of a state crime might achieve that goal. And given his previous experience, there was always the possibility he could escape again as he was being transferred to a local jail or in a courthouse.

There are a number of reasons violent offenders confess. If they already have been convicted of the crime, they might figure they have nothing to lose. Likewise, they might talk if they aren't getting out of prison in their natural lives and the crimes they talk about that have not been tried are not eligible for the death penalty. Some are just plain bored. And others want the perverse credit and recognition that comes from being regarded as a big-time criminal.

And this brings up an additional motive for Franklin's selective confessions. The simple way to describe it is: What good are your "accomplishments" if you don't get recognition for them? A serial killer who takes pride and personal satisfaction from his crimes is always going to be in an emotional bind. He can't let anyone know without being arrested, but as long as it's a secret, he's still a nobody. For criminals for whom the act of killing is the most important aspect of their lives—the aspect that gives their lives meaning—it is emotionally difficult not to be publicly associated with the murders.

Two of the most glaring cases of this in my career were New York City's Son of Sam and the BTK Strangler in Wichita, Kansas. It wasn't enough for seemingly mind-mannered postal clerk David Berkowitz to kill couples in cars in 1977 and then return to the scenes and relive the feeling of power and domination while he masturbated. He had to have recognition, which is why he styled himself the Son of Sam, acolyte of the three-thousand-year-old demon residing in his neighbor's Labrador retriever, and wrote to NYPD detective Captain Joseph Borelli and *Daily News* columnist Jimmy Breslin. It wasn't enough for municipal inspector, cop wannabe, and general loser Dennis Rader to carry out his cherished "projects" of breaking into homes, waiting for the occupants to return, tying them up in agonizing poses, strangling them, and watching them die; he had to let the police and media know that a serial killer was deserving of their attention and respect. Craving attention like Berkowitz, he also gave himself a moniker that appealed to him. Rader, living employed and undetected in Wichita, could not afford to reveal himself, though he flew as close to the flame of fame as he dared by sending letters to the media. And ultimately, it was a combination of written and digital communication that finally undid him. Were it not for that communication, he might never have been caught.

I don't think Franklin was as undisciplined as either Berkowitz or Rader, though he was just as irrational in a lot of ways. But though he saw himself as mission oriented, the publicity and recognition were important to him. How was he going to lead a race war if no one knew who the leader was? It doesn't matter whether the motive is sexual and "artistic" gratification, or a calling to foment a race war and become a hero of right-wing racist extremists. Franklin was already in prison, in all likelihood for the rest of his life, so the glory of the notoriety was a powerful lure. He had worked hard and devoted his life to these murders—in his mind, why shouldn't he get the credit and attention he deserved?

The motivation is even more complex. Our research had already revealed to us that three urges to power were operant in the minds of just about every serial killer: manipulation, domination, and control. Reaching out to law enforcement agencies and the media, having them respond to his beck and call, alternately confessing and then denying, and getting a chance to show off and assert himself in a public courtroom would fulfill all three urges, as well as getting him out of Marion and temporarily alleviating the boredom of maximum-security lockdown.

Ironically, Flynt, at the time, was serving his own fifteen-month sentence in the hospital wing of the Federal Correctional Complex in Butner, North Carolina, for contempt of court in refusing to disclose how he obtained videotape evidence in the cocaine smuggling and trafficking case against automobile entrepreneur John Z. DeLorean. And from prison, Flynt told CNN reporter Larry Woods that he had put out a contract on the life of President Ronald Reagan and would kill him himself, "if I can ever get anywhere close to him." Flynt was much fuller of bluster than Franklin, but, given my experience in law enforcement, I tend to take every threat seriously.

Gwinnett investigators still weren't sure how solid their case was because at the same time Franklin had confessed to them, he gave a long, rambling telephone interview to the *Cincinnati Enquirer* in which he stated, "I hate Flynt . . . because he publishes Black-white dating. But I had nothing to do with shooting him." By this point, my impression was that Franklin said whatever was in his head at the time, or whatever he thought would give him the easiest time going forward, with little or no regard to either truth or consistency.

And he was only beginning to talk. While Franklin was apparently willing to confess to crimes that he'd long been associated with, he was also willing to bring up crimes that had never been previously connected to him.

Late in 1983, an attorney representing Franklin contacted the police department in Madison, Wisconsin, and said his client wanted to talk about a murder in their jurisdiction in August 1977. By the time Detective Captain Richard Wallden was able to get in touch with Franklin the following February, Franklin told him of killing a Black man and white woman at a large shopping center and killing another woman in the spring of 1980 in Tomah, about midway between Madison and Eau Claire in east central Wisconsin.

With regard to the Tomah murder, Franklin said he had spotted a young woman near the East Towne Mall and offered her a ride. He said he thought she was on her way to Minnesota. It turned out she was actually on her way to her parents' hog farm in Frederic, Wisconsin, with a planned intermediary stop in Tomah to pick up her car, which she'd left at her uncle's house. In the car, he asked where she had gotten her suntan. She said she had recently returned from Jamaica. According to Franklin, she said she preferred Jamaican men to African American men because they were better-looking, and with that, her fate was sealed.

When they were close to Tomah, he asked if she wanted to smoke some marijuana. He pulled off the road near what he thought was a state park. It was isolated. He pulled out his .44 Magnum and ordered her out of the car, as if he intended to rape her. As she was walking away from the car, he cocked the handgun and shot her in the back. He left the body there and threw her belongings out of the car.

He claimed to make a habit of picking up female hitchhikers because he didn't think it was safe for them to be alone, but apparently, if they violated his own self-established rules of life, then they didn't deserve to live.

Also in 1980, Franklin said he picked up two white females in what he thought was Beckley, West Virginia. He recalled that they were on their way to some kind of anti-nuclear demonstration. He remembered the phrase "Rainbow Meeting" as well as them telling him they were Communists, which I doubted. It would be one thing for two young women to say they were anti-nuclear protesters. I doubted in 1980 they would identify themselves as Communists. I thought *Communist* was simply one of Franklin's all-purpose descriptions for people he did not approve of.

During the course of the conversation with the hitchhikers in West Virginia, he either asked whether or perceived that they went out with Black men. He said he shot them both with the same .44 with which he had killed the previous hitchhiker.

As appalling as this kind of crime was—killing white female hitchhikers upon their self-reporting of relationships with African American men—as revolting as I personally found it, it didn't really surprise me. Rather, it confirmed Franklin's obsession with race mixing and the erosion of white America. But there was another factor that spoke directly to the neurotic insecurities of Franklin and his ilk. One of the most powerful tropes in American racism is the fear of Black male sexuality. While it was common for white men in the antebellum South to rape their female slaves, the notion of Black men having sex with white women was about as abhorrent an idea as a racist male could conceive of. Though miscegenation was the stated concern, and though African Americans have fought the stereotype and depiction for more than a century, the subtext that racist men felt they were sexually inferior to Black men and therefore in danger of losing their women has been a motivating factor for much of the hate and violence visited upon African Americans throughout our history. As soon as I learned of Franklin's murders of hitchhiking women, I was certain this fear of inadequacy was combined with all of his other compensatory criminal behavior.

He told Wallden no more about the incident in West Virginia, but this would develop into one of the most perplexing and controversial cases attributed to Joseph Paul Franklin.

They spoke for a little over half an hour. After the call, Wallden telephoned Tomah PD, which confirmed that there had been an unsolved homicide in the area Franklin described, in which a white female was shot twice with a .44 Magnum. Her name was Rebecca Bergstrom, she was a twenty-year-old student, and the crime had taken place in Mill Bluff State Park. Her body was found the next day, May 3, by two teenagers and she was identified by the passport she was carrying. She was fully clothed and there was no sign of sexual assault. Since money was found in the wallet in her purse, robbery was also ruled out. A local newspaper article at the time said she had recently returned from a ten-day vacation in Jamaica and had a summer job waiting for her at a Frederic bank. She had been shot in the head and back, and spent shells were found nearby. There were no suspects, though Sheriff Ray Harris speculated she had been shot by someone who had offered her a ride. She was described by those who knew her best as friendly and loving.

Unexpected as these revelations were, the unsolved crime

Franklin was mainly calling Wisconsin law enforcement about concerned Alphonce Manning Jr., who was Black, and Toni Schwenn, who was white, both twenty-three years of age. They were shot to death at four thirty on the afternoon of August 7, 1977, in the parking lot of the East Towne Mall in Madison, near the JCPenney department store. After checking out his information, Captain Wallden sent Detectives Greg Reuter and Ted Mell to interview Franklin at Marion on February 16 and 17, 1984.

Investigators from Madison PD and the Intra-County Major Crimes Investigative Unit had only sketchy details to work with. There was a description of a white male in his early twenties with ear-length brown hair, driving a Chevrolet Impala with out-of-state license plates. The recalled color scheme of white with green numerals matched plates from Alabama, Idaho, Illinois, and Indiana. Witnesses said the car bumped the black Oldsmobile Toronado Manning was driving, at which point Manning stopped and got out of his car, whereupon the driver shot Manning several times as he approached. Then he got out of his car and went over to where Schwenn was sitting in the passenger seat and shot her through the car window as she was trying to escape, shattering the glass. The suspect was described as about five feet ten, 170 pounds, and wearing a dark green tank top and blue pants. A broad-brimmed hat found behind Manning's car was believed to belong to the killer.

Madison police officer Martin Micke was at the mall investigating a stolen car report and talking to the owners in his squad car. When he heard two or three loud gunshots, he had the couple immediately get out of the car and Micke sped to the scene and radioed for assistance. As he approached, one of the witnesses waved for attention and pointed to the green Impala racing out of the parking lot. Micke took off after it but lost it in a confusion of traffic and pedestrians. Responding to Micke's call, police set up roadblocks on highways leaving Madison but didn't find the Impala.

Schwenn died instantly at the scene. Manning died in a local hospital about an hour later.

According to the local *Capital Times* newspaper the following day, narcotics and vice squad officers were called to the scene to determine whether the broad daylight murders were drug related. This was routine procedure in daytime attacks in crowded areas, but the detectives didn't uncover anything that would lead them to suspect a drug deal. Detective Captain Stanley Davenport said there was nothing in either victim's background that would explain why they were murdered.

In an open letter published by the newspaper the following Wednesday, Toni's close friend Cathy Teegardin publicly speculated that the murders were the result of "racism" by a "maniac assassin."

I thought this must have been a particularly difficult case for the police emotionally because Schwenn had worked in law enforcement, as a typist-receptionist with the Dane County Juvenile Detention Center. For the past two years since he arrived in the city, Manning had been a janitor at Madison East High School, from which Schwenn had graduated five years before. She had been in the concert and dance bands.

Schwenn's mother, Janet, told *Capital Times* reporter Ed Bark, "To us she was the greatest. You can ask any of her friends."

Judy Manning described her older brother as a "nice person

who did not bother anybody." He had been a star football player on his high school team in Ruth, Mississippi. Friends said he was always eager to help anyone who needed it.

"He had a strong work ethic. He didn't sit back and let people take care of him," his cousin Tenia Jenkins-Stovall told a UPI reporter. He and Toni had met at a nightclub shortly after he arrived in Madison.

Toni's best friend, Linda Langlois, recalled, "She was real outgoing and physically beautiful. Alphonce was on the quiet side and Toni was loud and silly, so it was a good pair."

To me, the victimology has always been a critical part of any case, and I am often struck by the comparison in personalities between the victims and the offender. And here we had it again—warm, sensitive, life-loving individuals who had their young lives cut short by a sociopath who didn't even know their names, but could easily pull the trigger repeatedly because they were of different races from each other and were moving their car too slowly to get out of his way.

After being Mirandized, Franklin told Detectives Reuter and Mell he had originally gone to Madison with the intention of killing Dane County Circuit Court judge Archie Simonson, a Jewish judge who'd presided over a case of three Black men who had been accused of raping a white high school student and later freed. He told Reuter and Mell, "When I heard about that, I just decided to go up there and kill the bastard."

I found his plan fascinating as an insight into the criminal mind:

At first, what I was going to do was just going to walk up to the door—found out where he lived—first go to his chambers during the day and find his court, so I was pretty sure I wasn't getting the wrong person, ya know. I hate to kill an innocent person, ya know, so I figured I would just find out how he looked, would just walk up to his door one day, like on a weekend or something, when he would be home, and would have the pistol right back here, so I could fast draw it.

Among other things, this passage tells us that he is operating strictly within his own belief system and that he holds it higher than established law. He is the one who gets to decide who is innocent and who is guilty and what the rules of society should be.

If he couldn't shoot the judge, he had a backup plan: "I just got through bombing a Jewish synagogue down in Chattanooga, Tennessee, ya know, and plus, a Jew's home on the grounds [He seems to be referring to the rabbi's home] . . . I was also considering about wiring, I had five or six sticks of dynamite saved and some number eight blasting caps, was going to wire this Jew's car when he got in, ya know, if shooting wasn't very feasible."

What happened instead, in his account, was that he was driving toward the capitol building to get a look at Simonson before going to his house later on to shoot him when he saw two young women waiting at a bus stop. He picked them up and gave them a ride to East Towne Mall. As he was leaving the mall, another car pulled out of a parking space, blocking his way. It drove very slowly down the middle of the lane. Franklin leaned on his horn. The other car stopped, and the driver, who was Black, got out and starting walking toward Franklin's car. A white woman remained in the car. Since he had two guns and a load of dynamite in his trunk and thought he might have been spotted by a patrol car, he didn't want to get into a fight with the man and draw police attention. Then, as he sometimes did, he turned fatalistic. If he was going to prison, he would do so in service to his "mission." He opened his car door, shot the man, Alphonce Manning Jr., twice, then walked over to the car and fired two shots through the driver's-side window at the woman, Toni Schwenn, as she tried to get away. He recalled his black felt hat dropping to the ground as he went back to his car. He described the hat in such detail that the detectives were certain it was the one found at the crime scene.

Clearly, that pretty much matched the account witnesses had given police right after the shooting almost seven years before. Franklin described how he turned off East Washington Avenue onto Interstate 90, then turned off, returned to Madison, stopped at a McDonald's, then drove around until the police presence had diminished, and returned to the Ramada Inn where he had stayed the night before. He left town the next day.

As he was getting back on the interstate, he saw several police officers, but they apparently didn't see him. "Evidently God blinded their eyes," Franklin said. It later came out that the police were there to deal with a cattle truck accident and that they were consumed with helping injured motorists and rounding up cows.

This all seemed pretty credible to me. This was not the kind of murder in which he would get glory, so I didn't think it was made up or embellished. In fact, it was a failure of sorts, since it diverted him from his professed mission of assassinating Judge Simonson. While manipulation, domination, and control were still prime motivators for Franklin, he didn't seem to have much interest in lying when it came to revealing another crime.

Looking at this double murder within the larger timeline of his crimes, it was clear that these killings were a significant turning point. This crime was the one that opened his emotional floodgates, that truly unleashed Franklin's reign of terror. He had already threatened presidential candidate Jimmy Carter. He had also apparently bombed a synagogue. But on his way to murder a supposedly Jewish judge, he had accidentally run up against the personification of his hatred and mission and found himself spontaneously able to kill. Now, suddenly, he knew he could do it, and he could get away with it.

Every serial killer has a number of formative experiences along his pathway. This was a critical one for Franklin. He didn't think about it ahead of time; it wasn't a planned assassination. But the fact that he could take such "decisive" action on the spur of the moment would have given him the confidence, even if he only realized it subconsciously, that he could kill in any situation without hesitation. This, to a man like Franklin, would be liberating. It would represent the ultimate power.

Though Franklin was not a traditionally minded sexual serial killer, I could see here a trait he had in common with just about all of them: the depersonalization and objectification of his victims. They were not individual human beings to him; they were merely stereotypes of African American men and white women. This complete lack of empathy not only drove him, it actually allowed him to commit these senseless and outrageous crimes and then talk about them dispassionately years later. By fitting the stereotypes in his mind, the victims were no longer "innocent" and therefore the murders were justifiable.

At the same time that he depersonalized his victims, in typical paranoid fashion, he had an exalted view of his own importance and a detailed systematized belief in the machinations of organizations he perceived were out to get him personally:

> *I think they're trying to get rid of me, man. I just know too much about the government, international Communist-Jewish conspiracy, ya know, what the Jews are doing, all the right wing, all the Nazis. The Jews run every one of them, so they can just get the names of all the people who feel like I do, white racists and all, ya know, keep files on us, monitor activities, stuff like that. If you've ever been a member of it, they can automatically convict you of any charge, like if you've been with the Nazis, Ku Klux Klan or American Nazi Party, all they gotta do is get in front of that jury and say he is a member of this, and this is professional killing, therefore he is guilty, so they could, you may as well forget it if you go on trial or anything.*

He gave the detectives some printed pamphlets. "Read those pamphlets there I gave you there," he suggested.

"Yeah?" said Mell.

"There's a lot of truth in there."

"Okay," said Reuter.

"That's what's happening in the world."

CHAPTER 14

Franklin was getting a lot of official visitors. Shortly after his phone conversation with Captain Wallden and the two-day interview with Detectives Reuter and Mell, Ronald Pearson, chief deputy of the Monroe County, Wisconsin, sheriff's department, went down to Marion with Ernest V. Smith, an agent from the Wisconsin Department of Justice, to talk to Franklin himself about the murder of Rebecca Bergstrom, the lone female hitchhiker. Pearson and Smith came away with a confession and Pearson said he expected charges to be filed the following week.

When reporters asked him what the motive had been for the murder, he said, "Same as in Madison. It was a racist motive."

And sadly, police had been closer than anyone had previously realized in linking it back to Franklin. As would become clear after Franklin's confession to Wisconsin law enforcement, police had been on his trail. Just after the murders in the East Towne parking lot, two Madison PD detectives, Charles Lulling and Steven Urso, narrowed the plates to Alabama and went to Montgomery to search motor vehicle records. After going through thousands of vehicle records, they were down to fifty-five people who owned 1967 Chevrolets and lived in the Mobile area. They intended to contact and interview all of them, but after they'd talked to about fifteen people, they were called back to Madison because police officials didn't think they were making any progress.

Lulling and Urso disagreed vehemently. They were convinced that the shooter was among those fifty-five and that as soon as they talked to him, they would know it.

"All we needed was a confrontation with him and we would have scored," Lulling, by then the head of a private detective agency, told the *Wisconsin State Journal* in 1984. "There's absolutely no doubt in my mind."

At a press conference on Friday, March 2, 1984, Madison police chief David Couper disputed Lulling's evaluation, saying that Detective Captain Stanley Davenport had ordered them back home because they had not produced any specific leads.

Either way, Joseph Paul Franklin went on to kill at least nineteen more human beings.

IN THE WAKE OF FRANKLIN'S CONFESSION TO HIS WISCONSIN KILLINGS, HIS visitor list at Marion kept growing because he kept talking.

One of the crimes Franklin had mentioned during his initial two-day interview with Reuter and Mell was the bombing of a synagogue in Chattanooga, Tennessee, on the night of July 29, 1977, which law enforcement had never connected with him. The blast had leveled the Beth Shalom Synagogue and scattered debris for a block around the wood and brick building. It shattered windows in a nearby motel and woke Rabbi Meir Stimler, who was asleep in his home behind the synagogue. According to the Associated Press, "Witnesses said a section of the roof was blown into the air and landed virtually intact 20 yards in front of the demolished structure." Because of the time the bomb went off, no one was killed or injured.

At first, apparently unable to conceive that someone would intentionally blow up a house of worship, Rabbi Stimler discounted the possibility of a bomb having caused the explosion, which left a crater two feet deep near the middle of the building.

Police bomb squad experts had a different take. When they examined the wreckage, they found evidence of nitrate deposits beneath what had been the floorboards, and an electrical cord used to detonate an explosive device. A long electrical extension cord apparently had been run to a socket on the exterior wall of the Airport Inn motel, about two hundred feet away. A police explosives expert said the bomb was "high-grade and highly sophisticated." A neighbor said that when he felt his house shake he thought an airplane had crashed.

Still, more than six years later, the case remained unsolved. And when discussing the Wisconsin crimes, Franklin brought up his knowledge of explosives—a skill, like bank robbing, that he'd honed during his criminal career—which led him to recount the times he'd used bombs to try and kill people.

"Not really hard," Franklin related. "But if you don't know anything about explosives, you can get yourself blown up to pieces, and I put fifty pounds of dynamite in there, ya know, right underneath the synagogue, and wired a line all the way through the grass, next to the hotel. So, I found out there were no Jews in it, so I just plugged the sucker in, and it just went

Choom! And it just disintegrated, man. It made big news, ya know. Matter of fact, I read about it when I was up in Ohio."

The crime had gone into ATF's cold case file and been there ever since, but perhaps sharing the story with Madison Detectives Reuter and Mell made Franklin more willing to talk, because once again, he called the police to come talk to him at Marion. On February 29, 1984, less than two weeks after Reuter and Mell had been there to interview him, ATF agent George Bradley and Chattanooga police inspector Charles Love arrived. Wisconsin investigators Pearson and Smith would be next in line after that.

Their first item of business was to read him his Miranda rights while recording the conversation. "Are you willing to make a statement to us at this time without a lawyer present?" Love asked.

"Yes, I am."

"This initial interview will concern bombing of the Beth Shalom Synagogue."

"Right."

The motive for the confession appeared to be the same as in the Flynt shooting—a combination of wanting credit, relieving boredom, and perhaps even some kind of personal inventory, now that he had nothing to do but think about his past life. Meticulously, the investigators went through Franklin's planning—where he had purchased the dynamite and how much he had acquired; how he had surveilled the site at night and found a crawl space under the building to place the pack of fifty pounds of water gel explosives and five sticks of dynamite; how he had looked around for a place to plug into an electrical outlet to detonate the dynamite; how he had run the electrical cord from the nearby motel; and finally, how he had called the synagogue to ask when they would next "meet," "making out like I was interested in going to one of their meetings." He confirmed everything the investigators had determined.

"I went back there on the day that they said they were supposed to be there and, uh, hooked another extension cord from the end of the cord that I had there to the motel."

"Did you stay around and watch the results after you detonated it?"

"Naw, I just figured I'd split as soon as possible."

Love showed Franklin photos taken of the synagogue after the explosion, including one taken from the air, and asked him to identify surrounding buildings. Franklin seemed quite interested in studying the shot. "Man, was that it right after?" he asked.

"That's correct," Love replied.

"Man! Blew that tree down, too, huh? That sucker really got pulverized." Clearly, he was highly impressed with his own work.

This dimension of Franklin's criminal personality was very interesting to me in terms of our analysis of both modus operandi and signature. Normally, the means of killing is either an inherent part of the signature, as strangling and mental torture were to Dennis Rader and physical torture and terror were to such perverted sadists as Lawrence Bittaker and Roy Norris, and Leonard Lake and Charles Ng; or it is an M.O., as it was with Ted Kaczynski with bomb-making. That is to say, killers kill through the means they find most comfortable. As a rule, a shooter remains a shooter, a bomber remains a bomber, and a strangler remains a strangler. There may be evolution in the seriousness of the crimes, like a Peeping Tom evolving into a rapist or a fire-starter like David Berkowitz evolving into a killer. But they generally tend to remain with their preferred form of violence.

Franklin was different. He didn't seem to get off on the act itself, but on the results, suiting the means to the situation. In his case, it was the signature that was all-important: killing African Americans, Jews, and those who associated with them. But the M.O. was fluid and adaptable. Whatever it took to rob a bank successfully, he would do, and learn from each experience to be better at it the next time, just as a baseball player might adjust his stance or swing to improve his batting average. He had begun his mission by bombing a private home in Maryland and a synagogue in Chattanooga, but much to his disappointment, neither act resulted in any human fatalities. At that point, I believe, he decided his efforts had to be more direct. He was at his best as a sniper, because that is the skill he had perfected after the eye accident of his youth, so that was the course he followed. Ironically, it is as if this extremist right-wing high school dropout had unconsciously adopted the slogan popularized by such left-wing intellectuals as Frantz Fanon, Jean-Paul Sartre, and Malcolm X: *By any means necessary.*

"How long were you in Chattanooga before you blew up the synagogue?" Love asked.

"Uh, let's see. Well, actually, I was there, I bought the explosives and then I left, you know, for a while. And then came back there." And here, surprisingly, Franklin segued into yet another bombing confession. "You know, at first I wanted to, after I bought the explosives from the store in Chattanooga, then

I went up to Maryland and decided to bomb the Jew's home up in Maryland, you know, while I was up there."

"Now, whose home was it in Maryland that you bombed?" the investigators asked.

"A Jew by the name of Morris Amitay: Morris A-M-I-T-A-Y."

"And how did you select him?"

"Uh, I just happened to be reading the *Washington Post* one day, and it said something about an Israeli lobbyist was doing something there, you know. . . . So, uh, being familiar with Rockville and Silver Spring, Maryland, I had lived there for so long, I knew exactly where the area was. So, I went over there and looked up his name . . . and just went over there, cruised, you know, passed by his house. He lived right on the corner, and uh, checked it out a couple of times. Then I went back there later, one night, and blew it up."

Four days before the Beth Shalom bomb, the home of attorney and American Israel Public Affairs Committee (AIPAC) executive director Morris J. Amitay in the Flower Valley neighborhood of Rockville, Maryland, a D.C. suburb, had been rocked by an explosion. The split-level house was severely damaged in the 3:20 A.M. blast, and though Amitay, his wife, Sybil, their two sons, and a daughter managed to escape unhurt, they lost their six-month-old beagle Ringo, who was in the ground-floor family room. Windows were broken and siding torn loose in houses as far as five blocks away. "I don't know how anyone got out of there alive," one fireman commented.

The forty-one-year-old Amitay was a former foreign service officer and foreign policy and legislative aide to Abraham Ribicoff, the former health, education, and welfare secretary under President Kennedy and later Connecticut senator. Amitay said the family had not been threatened and he had no idea who could have committed the crime, though police speculated the attack had to do with Amitay's being regarded as one of the leading lobbyists for Israel in Washington. Investigators said they were unsure whether the explosion was meant to kill the family or merely frighten them.

"Nobody's called to take credit for it," said police spokesman Philip Caswell at the time. It is not unusual in terrorist crimes for the responsible group or others with similar aims to publicly claim responsibility.

I. L. Kenen, AIPAC's honorary chairman, said he thought those responsible were "anti-Israel, either anti-Semite or pro-Arab."

It is interesting to me that in those days, it was assumed a crime of this magnitude must have been carried out by a group rather than a single individual. Burton M. Joseph, national chairman of the Anti-Defamation League of B'nai B'rith, expressed shock and called upon the FBI and police to pursue the investigation with "vigorous dispatch so that the terrorist culprits are apprehended as quickly as possible."

Police and ATF agents determined the explosion was caused by dynamite and detonated with a four-hundred-foot electrical cord. Within a week, federal officials had begun coordinating investigations into the Chattanooga and Rockville bombs through the Bureau of Alcohol, Tobacco and Firearms' Explosives Enforcement Branch. An ATF spokesman said that after a blast, "A computer scans all our bomb cases and flags similarities. Then we immediately begin working on the two together."

Franklin drew a diagram of the Amitay house and described to Bradley and Love how he had determined where to place the explosive device:

> *I suspected that possibly since this corner was furtherest away from the street, you know, from the street noise, that he would, the bedroom of the house would be right there, you know? So, therefore, I decided to go ahead and bomb this corner right here, you know, so it would kill more of the, you know, when it hit.... So, I found out later I was mistaken.... It didn't kill none, you know. It blew their dog up.*

When they finished asking him about specific crimes, Bradley clarified, "As far as talking to us and telling us what you've done here, what you know, we haven't promised you anything as to what we could do in Tennessee relating to this crime."

Franklin replied that he just wanted to get a transfer out of the federal penitentiary, but that he really didn't expect anything in return. Bradley reminded him, "You know you're confessing to a serious crime."

"I don't really think it's serious," Franklin replied. "I don't even consider it a crime. I think it was good."

When Love asked why he picked the Chattanooga synagogue for bombing rather than New York, Franklin said, "I was basically familiar with Chattanooga and I was able to buy the dynamite there." I found this an interesting statement because it showed that even a mission-oriented criminal like Franklin still committed his crimes within a basic comfort zone.

After denying involvement with the Larry Flynt shooting, he admitted onetime membership in all the extreme right-wing organizations we had documented, but then explained why he dropped out. "They're all controlled by Jews and queers, you know, and FBI informants. And actually, the ADL [Anti-Defamation League] runs those." While this is a somewhat different explanation than he had given previously for why he had left these hate groups, the response is consistent in its contempt for his former comrades. More important, I think it demonstrates Franklin's growing paranoia.

His next response confirmed that paranoia. Love asked, "Did anyone aid you in any way? Did you let anybody in on what you were going to do? Or did anyone encourage you to do it?"

"Anyone in on it? No, uh-uh. I never told nobody about anything. The only ones that knew about that were me and God. That's the way I've always been ever since. I've never told nobody anything about what I've done. Not one single soul would I trust, you know. So, that's one reason I was able to operate so long without anybody, you, being able to catch me before. I never told nobody nothing."

"In case after case," the AP reported, "investigators say Franklin has been able to provide scraps of key information that could not be known by anyone who was not at the crime scene."

CHAPTER 15

In a matter of weeks Franklin had confessed to a startling array of crimes, and of these new confessions, Tennessee reacted the fastest.

On July 12, after a two-day trial, it only took a Chattanooga Criminal Court jury about forty-five minutes to find Franklin guilty of the 1977 synagogue bombing. There was a clear forensic trail of purchasing explosives from a Chattanooga supply store under his own name and fingerprints on the transaction records that matched his exemplars. The two prominent local attorneys who represented him were out to convince the jury that the taped confession was just another of Franklin's many lies, intended to get him out of Marion. It was an odd courtroom dynamic, with the prosecution trying to convince the jury the defendant was telling the truth, while the defense tried to show he was lying.

Though he refused to take the stand, the attorneys' strategy kind of fell through the floor after prosecutor Stanley Lanzo suggested Franklin was a coward in spite of his macho self-image. Telling the jury that Franklin was in a protective unit at Marion because of what the Black inmates thought of him, Lanzo said, "He ought to be in the general population where they can test him to see if he's a real man."

That was too much for Franklin. He asked Judge Douglas Meyer if he could make a statement in rebuttal. The judge granted the request, telling the jury that the defendant had decided to make his own closing statement. His attorneys were not exactly pleased, and I have to say his oration was compelling, at least in terms of deciding the trial's outcome.

"Jews control the American government," Franklin asserted to the eight-man, four-woman jury, which included two African Americans. "They control the news media, they control the Communist governments, they control the Western democracies. I'll admit to you I bombed the synagogue—I did it. It was a synagogue of Satan.

"This country was founded by white men who were believers in Jesus Christ. They've been taken over by atheists. I'd just like to tell you this—the only way for the white man to survive is to get on their knees and pray to the Lord and accept Jesus Christ as their personal savior."

Now, these statements pretty much speak for themselves, and we can easily see how they "helped" the jury reach its collective verdict. But for a profiler and criminal analyst like me, they provided some interesting insight into Franklin's personality. First of all, he is not making this speech to piss off the jurors or turn them against him. Rather, he is stating something he believes and wants to convince them of. He doesn't expect the jury to let him off, necessarily, particularly the two Black members; he's already in prison for a long, long time regardless of how it rules. But he does want them to understand why he did what he did and why he's proud of it. He wants his statement to get out—if the Jewish-controlled media will let it out—to inspire others to join his cause. So, there is a delicate psychic balancing here between outlandish paranoid fantasy and logical thinking about how to get his message across.

Second, it is clear that within his paranoid logic system, he truly believes he is a devout Christian and believer. Would Jesus want his followers to go out and hunt down Jews and Black people? No rational person would think so, and yet he has convinced himself that he is on a holy mission ordained by God and that Jesus is telling him what to do. In effect, he has psychologically inoculated himself against the charge that what he has done is wrong or evil, so he need not have any remorse or regret about any of it. It is the same mechanism by which sexually motivated serial killers depersonalize their victims. It stems from a complete absence of empathy and the narcissistic idea that they are the only ones who matter.

Are they insane? No, not by any legal definition. They are just really bad people.

Judge Meyer sentenced Franklin to fifteen to twenty-one years for the bombing and an additional six to ten years for possession of explosives. The terms were to be served consecutively, though everyone in the courtroom knew the sentence was essentially symbolic, since it couldn't be served until Franklin had served his life sentences for the Salt Lake City murders.

But the symbolism was important to the judge. "Jews have been persecuted for two thousand years," he said. "But regardless of what's going on in the Middle East and in Ireland, the word must go out that we will not tolerate crimes against humanity."

BY THE TIME FRANKLIN'S TRIAL FOR THE MADISON MURDERS BEGAN IN DANE County Circuit Court on Monday, February 10, 1986, he had recanted his prison confession to the two police officers. The previous month, Judge William D. Byrne had rejected Franklin's request that the confession be suppressed. He did, however, allow Franklin to act as his own attorney, with a lawyer and former state public defender, William Olson, there to aid him.

In bringing the Manning-Schwenn murder case to trial in Dane County Circuit Court, District Attorney Hal Harlowe explained that under Franklin's current sentences, he could possibly be eligible for parole as early as 1990, and he was pursuing further murder convictions to assure that he would never be released. "To my amazement, there really weren't any guarantees he would never be released."

I seldom like to make comparisons of real killers to fictional characters, but in one way—not intelligence—Franklin was similar to novelist Thomas Harris's Hannibal Lecter, who was also supposedly in prison for life, but everyone was still scared of him and what would likely happen if he ever got out. Franklin had repeatedly proven himself to be a hate-filled, efficient killing machine.

Acting as his own counsel, Franklin insisted to another all-white jury that he had faked his confession to get out of Marion and that he had gotten the details of the crime from the newspapers. "I just wanted to win temporary release from the federal prison in Marion, Illinois, because of the brutal conditions there."

Then, referring to himself in the third person he said, "The evidence in this case will show that the defendant was not the perpetrator of this crime. They have no case here other than a confession, and the defendant is now saying this confession is false." He further claimed that he was tortured into making the confession, which was patently absurd to anyone who heard the tape.

There was high security in the courtroom and Franklin's legs were shackled, though the jury couldn't see this as he sat behind the curtained defense table. Harlowe played the confession tape, in which Franklin described coming to town under the alias John Wesley Hardin, after the famous Old West gunslinger and gambler, with the intention of assassinating Judge Simonson but then ended up shooting the young couple in the mall parking lot.

Johanna Karen Thompson, a teller at the Ohio State Bank in Columbus, identified Franklin as the man who had pointed a .357 Magnum revolver in her face on August 2, 1977, as he robbed her bank of twenty-five hundred dollars. She wept softly as she pointed him out and said, "It was him." He had described the robbery in his confession and characterized it as "confiscating the money from the Jewish bankers." Richard Thompson, a ballistics expert formerly with the Wisconsin State Crime Laboratory (and not related to bank teller Thompson), affirmed that the .38-caliber bullets removed from Manning and Schwenn's bodies could have been fired by such a weapon.

The strong case built from there. Madison police captain Richard Wallden testified about his February 7, 1984, telephone conversation with Franklin. And then there was the confession tape itself—the disturbing and clear catalyst for the long-sought justice. As it played on the fourth day of the trial, the twelve jurors and two alternates heard Franklin's confession and his hate-filled words.

In his summation the morning of Friday, February 15, Harlowe called the murders "a cold-blooded execution" and said, "He is guilty of a horrible, senseless, pointless, unnecessary crime" that was the "closest thing to killing for sport" Harlowe had ever seen. He said, "He was proud of what he did. He wanted to talk about it. It was bottled up too long," and referred to the taped confession as "the pride of recollection from a job well done." My thoughts exactly. "Justice has waited nine years," he said. "I'm asking that it not wait any longer."

The confession was clearly believable to the jurors, who took only two hours to return a guilty verdict on two counts of first-degree murder. "The defendant's history of violence, terror, and murder prompts this court to do all it can so that he will never kill again," Judge Byrne stated as he pronounced the two consecutive life sentences that were mandatory under Wisconsin law. When he asked if Franklin had anything he wished to say, Franklin, showing no emotion as the sentence was passed, replied, "No, Your Honor."

Afterward, standby defense counsel Olson made several comments reported by Sunny Schubert for the *Wisconsin State Journal* that I found highly astute. He said he had once come to see Franklin in jail unexpectedly, which really threw Franklin off. "He was quite upset with me. I figured, the guy's in jail, what does he have to do? But he was angry because I upset his schedule," which Olson reported consisted of meditating, praying, exercise, and reading the Bible. Saying he was "wily," with "the skills of a hunter," Olson noted, "He's not good at abstract thinking. We all had fantasies when we were nine, ten, of being Superman or the Lone Ranger. Franklin never grew out of his."

That was all true, but you couldn't deny that Franklin was criminally sophisticated. In spite of a scattering of eyewitnesses who thought they had seen him in motels or the vicinity of shootings that had just taken place, in all the killings attributed to him or that he was suspected of, no one had actually seen him holding a weapon in the firing position, pulling a trigger, or committing an act of murder. Had it not been for the fact, as Hal Harlowe suggested, that "He wanted to talk about it. It was bottled up too long," this case would not have been brought to trial. So, on two levels, both the original acts and the compulsion to talk about them, Franklin had sealed his own fate.

Outside the courtroom, Harlowe commented, "I am opposed to capital punishment, but Mr. Franklin puts those beliefs to a sore test. Mr. Franklin is a pathetic creature who will be dangerous until the day he dies."

CHAPTER 16

Though Joseph Paul Franklin's list of crimes had grown substantially since I'd put together my initial profile of him, the review of his time behind bars had prepared me to meet him face-to-face for the first time.

As part of our joint Secret Service–FBI study project on the assassin personality, Secret Service special agent Ken Baker and I signed out a Bureau car and drove from Quantico to the U.S. Penitentiary in Marion, Illinois. Ken and I met originally when he was assigned by the Secret Service to attend the FBI's Police Fellowship program at Quantico. This was a nine-to-twelve-month training course in criminal investigative analysis, including interview and interrogation techniques, investigative and proactive methods, prosecutorial strategies, and expert witness testimony. Ken was an exceptional agent with an impressive Secret Service background, and after the completion of the Police Fellowship, he was permanently detailed to our unit. We also had an Alcohol, Tobacco and Firearms special agent, Gus Gary, working with us, and those two dramatically increased our scope and expertise.

My involvement with the Secret Service had begun in 1982, two years after I did the fugitive assessment on Franklin, when a Secret Service supervisor contacted me about doing an assessment on a character who signed himself C.A.T., who had written a threatening letter to President Carter, and now had written a series of them to President Reagan. The question the Service most cared about was whether this guy was actually dangerous, since one letter included photographs that he had been able to take close-up with a New York senator and congressman. Also worrisome was that the letters were postmarked from around the country, suggesting that C.A.T. was highly mobile, like Franklin.

After I constructed a profile, we placed a carefully worded ad in the *New York Post,* which C.A.T. answered, thinking he was responding to a newspaper editor, whom he was encouraged to call to arrange a secret meeting. The "editor" was, in fact, a Secret Service agent, whom we coached on how to speak with the UNSUB on the phone and draw him out. I thought he would call from a pay phone in some public place, like Grand Central or Pennsylvania Station or one of the large libraries. We put a trap and trace on the phone line and kept him on the phone long enough for a combined Secret Service–FBI team to locate and pick him up at a phone booth in Penn Station. Alphonse Amodio Jr. was a twenty-seven-year-old native New Yorker who had a grudge against the world for not paying any attention to him. He had no particular political agenda. He was institutionalized after his trial judge requested a psychiatric evaluation. I didn't think he was immediately dangerous, but what you always worry about with these types is when the strains of life become greater than the fear of death.

Our first stop on this trip, though, was to see someone far more mission oriented. On the way to Marion, Ken and I took a slight dip southward to stop at Brushy Mountain State Penitentiary in Morgan County, Tennessee, in the northern part of the state. Our subject there was James Earl Ray, who had assassinated Dr. Martin Luther King Jr., sniper style, in Memphis, on April 4, 1968. I wish I could say we got a lot of insight from him, but by this time, more than twenty years after Dr. King's murder, Ray was so bound up in his own paranoid fantasies that we couldn't be sure what he actually remembered, and we certainly couldn't get into his headspace regarding the time immediately before, during, and after the assassination.

Generally, we had found, the assassin-type personality is delusional and paranoid to some extent, but not in the sense of being insane. They are personally angry at specific types or groups of people, and anyone whose belief system is different from their own. But Ray was at the extreme. Though he had originally pleaded guilty, he now had recanted his confession and said he realized he was the unwitting dupe of a complex conspiracy to kill the civil rights icon. None of this rang true, but I can't say we were able to get any valuable understanding or useful information out of him. Not all of the prison interviews go the way you hope.

Likewise, I had no idea how the interview with Franklin would go or how he would react to us, so as we drove up through Kentucky on our way to Illinois, Ken and I talked out a strategy. We had already extensively researched his criminal file and read every newspaper story we could find. The approach I suggested to Ken was that we present ourselves as special agents from two different law enforcement agencies who were in awe of his criminal "accomplishments" and wanted to learn how he was able to outsmart us for so long. We decided it would be best if we wore our dark suits rather than the casual clothing that we normally wore when we interviewed serial killers, in an attempt to put them somewhat at ease and symbolically narrow the distance between us and them. With Franklin, we thought a different approach was called for. While our formal appearance would reflect our authority, we would let him take the lead in the interview process. We would also encourage him to ask us questions or suggest topics if he wished. This would feed his fragile but sizable ego and let him feel that he had the edge on us and was once again in control.

Interviewing assassin types is somewhat different from interviewing serial killers, rapists, or other violent offenders. They would almost always define themselves as mission oriented, though that mission might be political, social, or deeply personal. Though David Berkowitz, for example, had no sexual contact with any of his victims, his crimes were certainly a compensation for his own sense of sexual inadequacy. In addition to the motive of killing couples who had found relationships that he had not, his mission was to achieve fame and notoriety. He derived a sense of power from creating mortal fear throughout the city of New York.

We always tried to make the most efficient use of our travel time when conducting the prison interviews, so Franklin was not the only apparently politically motivated killer that Ken and I were planning on speaking with at Marion. We also wanted to meet Garrett Brock "Gary" Trapnell, who was serving a life sentence in federal prison for air piracy, kidnapping, and armed robbery.

Together, Trapnell and Franklin made for a study in contrasts in almost every way, except for the fact that both were very good at robbing banks. Trapnell was one of the smartest and, dare I say, the most clever and "charming" offenders I have ever dealt with, as well as the most resourceful. He had made his living as a con man, bank robber, and burglar. He robbed a string of banks in Canada, stole about a hundred thousand dollars' worth of jewelry in the Bahamas, took on a variety of pseudonyms and disguises, and carried on marriages with at least six women at the same time. In 1978, he was able to convince a forty-three-year-old woman friend, Barbara Ann Oswald, to hijack a helicopter in St. Louis and get the pilot to fly to Marion, land in the prison yard, and rescue him, perhaps the boldest jailbreak idea of all time. During the landing, pilot Allen Barklage, a Vietnam veteran, managed to wrestle a gun away from Oswald and kill her, thereby thwarting the plan.

His most notorious crime was what got me interested in him as a subject. On January 28, 1972, he boarded TWA Flight 2 in Los Angeles with a .45-caliber handgun hidden inside a plaster cast on his arm. Midway through the flight to New York, he hijacked the plane and made a list of demands: $306,800 in cash (he had recently lost that amount in a court case), the release of imprisoned Black professor and political activist Angela Davis, and a formal pardon for himself from President Richard Nixon.

When the plane landed at Kennedy Airport in New York, Trapnell released the ninety-three passengers but held the crew at gunpoint so he could negotiate. He threatened to crash the plane into the terminal if his demands were not met. After about eight hours, he agreed to allow a crew switch and refueling. During the switch, an FBI team dressed as ground crew boarded the aircraft and managed to shoot Trapnell in the arm. As he was being led off the plane, he again repeated his demand regarding Angela Davis.

His first trial ended in a hung jury when one of the jurors bought his insanity defense, based on a multiple personality claim. He second trial resulted in conviction.

As we neared Marion, Ken and I expected to confront two politically minded criminals on opposite sides of the political spectrum. Trapnell took a huge risk in trying to get Dr. Davis freed, even greater than Franklin took every time he killed a Black victim. But I couldn't figure out the reason for Trapnell's demand. Nothing I could find in his background indicated any particular connection to left-wing, civil rights, or radical causes. There were rumors that he had a romantic obsession with Davis, but that would have been out of character for him. So, what was his actual motive? We wanted to see whether Trapnell and Franklin were actually similar characters with opposite missions. If so, that would give us important insight into the mission-oriented criminal personality.

Marion is a large, bland complex of buildings enclosed by two parallel rows of fencing and surrounded by a perimeter road. It is in the midst of a green field cut out of the surrounding forest. The access, appropriately, is from Prison Road. We pulled into the parking lot outside the main gate and then checked in. It was funny to look at a row of buildings, imagining the prison yard behind them, and wonder how many ways both Trapnell and Franklin had thought of to try to escape from its confines.

CHAPTER 17

When we met with Franklin, he was still being held in the protective basement K Unit. He came up from the cellblock by a staircase with a metal railing that Ken and I could see through the door of the small room where we were waiting. You could hear the jailhouse chatter from down below. Word had gotten out that two FBI agents were in the prison. We were surprised that Franklin came up on his own, unaccompanied by a guard.

Our meeting room had beige walls, windows covered with bars, a table, and plastic stacking chairs around it. Ken and I were wearing our darks suits. Franklin was wearing thick glasses, jeans, and a blue prison shirt. I remember that his light brown hair, which varied in length over the years, was long and wild. He seemed quite upbeat and gratified that two federal agents were there to see him. We indicated a chair, but he remained standing, as he would throughout the interview.

We were not there to interrogate him or clear any cases, though I would be happy if we had. We were there to learn more about how his mind worked, and how we could apply it to our study of assassins, snipers, and similar types of offender. If you think it's like a cross-examination in a murder trial, it's not. If you think it's like Clarice Starling's confrontation with Hannibal Lecter, it's not. The idea is to keep it calm and easygoing. You don't want to get confrontational or in his face, because you don't want him to manufacture any false memories because you've gotten him riled up. That would be counterproductive to the objective, which is to get a sense of his actual feelings and motivations and how they correlate to the crime itself. In most cases, we are dealing with what German Jewish philosopher Hannah Arendt, reporting on the 1961 trial of Adolf Eichmann in Jerusalem, called "the banality of evil." Despite how emotionally satisfying it might be on a personal level, I've never been accusatory or expressed outrage or moral superiority in a prison interview; if it gets too tense from the subject's perspective, then I will not achieve the objective.

We had several goals, among them to see how closely Franklin matched my initial evaluation of him and to get a stronger sense of his actual motivations. Nine years before, in 1980, when I did my fugitive assessment, profiling and behavioral analysis were still new and experimental. Frankly, I was guessing at a lot of what I came up with. Now, with the program firmly established and an expert team working with me, I wanted to see how right or wrong I was on various pieces of the total picture.

We knew that before Franklin set out on his murderous rampage, he had separated himself from the various right-wing groups he had joined when he decided they were more talk than action. But one of the things we were really hoping to determine was where and when his movement toward dangerousness actually began. What were the precipitating events in his life? How organized was he, actually? How much planning did he do? And what was going through his mind as he committed each crime, and in the hours and days afterward? As far as I have been able to learn, he had never claimed responsibility for a crime he didn't actually commit, so I was at least confident that we would get reliable factual information from him.

I can't give a blow-by-blow account of our conversation with Franklin because after our first interview, with Ed Kemper, we stopped recording them. Unlike an interrogation, where you are looking to substantiate facts and disprove lies, the prison interviews were intended to be more subjective and expansive, to focus on emotions and sensations, and we wanted our subjects to think and speak freely, unencumbered by the concern that their words would be used against them. This was particularly important for paranoid types like Franklin.

No matter what they're in for or how long their sentences may be, most convicts entertain some vague hope of one day getting out, whether to go straight or resume their criminal careers. So I would usually begin with some kind of encouragement to cooperate, like, "We can't guarantee anything, but by participating in this, you will be helping law enforcement with our understanding, and we will make it clear to the warden and other authorities that you answered our questions honestly and forthrightly." This can be particularly effective with someone like Franklin; we knew from studying his background that he had once wanted to be a cop, so we tried to make him feel like a "partner" in this exercise.

Through my words and body language, I tried to make it

clear that our approach was to be agreeable and nonjudgmental. Just as in hostage negotiation, the aim is to listen to what the subject is saying, then restate the content and paraphrase it so he knows you're both on the same level. If you let them hear back what you think they're telling you, that can go a long way toward establishing trust and making sure you get the message clearly.

Franklin was polite and almost affable. On one level, this didn't surprise me. If he were sullen and withdrawn or overtly threatening when dealing with "challenging" interpersonal encounters, no hitchhiker would have ever gotten into a car with him. At the same time, he was wary, and his eyes kept darting back and forth between Ken and me, trying to size us up. This was in contrast to the transcripts I'd gone over of his sessions with the police detectives and ATF agent, where he'd been forthcoming right from the beginning. I think the difference was that in those instances, he had initiated the meeting, so he felt in control. Since we had asked to speak with him, he didn't know what our agenda was.

Still standing, he asked me in a joking way if I was one of the FBI agents who had infiltrated the Ku Klux Klan, because he said the Klan probably had more FBI informants in it than actual members.

Once we made it clear that we had studied his background and rap sheet extensively, he started opening up more. We explained our assassin study and said that since he had been such a successful and prolific killer, we thought it was important to include him. He nodded and seemed to appreciate this validation. This tactic tends to work best with the more self-important killers, the ones who take themselves most seriously, which was certainly the case with Franklin.

My technique in the prison interviews was to tell the subject about himself as I understood him and see how he reacted. This tended to compliment the subject that I had spent so much time and effort studying his life, and it also would get him to react in such a way that he would give a good sense of his own self-image.

I reviewed his early life and his relationship with his parents. I said it was my understanding that he had started out as a good student but had lost interest in school. He was not much interested in athletic teams or extracurricular activities and pretty much kept to himself. He listened without saying much, which let me know we were on the right track.

To see how far we could take the assassin aspect, we started off by asking him if he had been aware that President Carter was going to be making a campaign stop near him in Florida in October 1980. He admitted he knew about it. But when we asked him, given the letter he'd written to the then-candidate, whether he had planned an assassination attempt, he seemed to slough off the possibility. Not that he wouldn't have minded shooting the president from a distance, but he said, like Arthur Bremer, that since the assassination of President Kennedy, it had been too difficult to knock off a president, and that he certainly didn't think he could have pulled it off from a sniper's nest. He said he didn't consider President Carter important enough to sacrifice his own life, even if he had been able to get close enough with a gun.

That didn't mean he had backed off any of his racist and anti-Semitic philosophy. Just as with the previous interviews he'd had with investigators, he was completely matter-of-fact with his views. His only regret was that he hadn't been able to start a full-scale race war, though he knew he had inspired other like-minded individuals.

What struck us both was how sincere Franklin was about his beliefs, how unashamed he was, and how little he cared whether he was popular or what the general public thought of him. This is not to say he was indifferent to the response of people who thought the same way he did; he definitely saw himself as a hero to that cohort and wanted their approval. More important, he made clear that he felt he was in essence writing a script for them to follow. That is fairly unusual for a repeat killer. Either they are only interested in fulfilling their own dark fantasies and getting away with each crime, or, like David Berkowitz, they want the wide-scale notoriety and sense of power that instilling fear gives them. Franklin had a clear-cut distinction in his own mind about who he wanted to influence. I would therefore have to characterize him as thoughtful, at least about the subjects that interested him.

I asked him about the Vernon Jordan attack—the one crime he had never admitted to—setting up the scenario in Indiana and describing how I thought it went down. In response, he smiled like the Cheshire Cat, and with an expression I can only characterize as pride asked, "What do you think?" I shrugged, indicating he should continue. "I'll just say that justice was done," he said. I detected a real internal conflict here. When Ken and I talked about it afterward, we strongly believed Franklin was afraid that if he admitted that one, word would get out in the prison and the Black inmates would figure out a way to get at him.

On the other hand, he freely admitted shooting Larry Flynt for publishing features with mixed-race couples. He made it clear that despite what he considered his strongly Christian beliefs, he was not against pornography; he was actually quite an avid consumer. But when he saw the pictorial involving the interracial couple, he blew up.

Originally, he had gone to *Hustler* headquarters in Columbus, Ohio, where he looked up Flynt's home and office addresses in the local telephone directory. But when he got there, he said, he discovered Flynt was in Georgia for the obscenity trial and was staying with President Jimmy Carter's evangelist sister, Ruth Carter Stapleton. We had checked this out before the interview and it is true that the two of them had a relationship and Stapleton even converted the pornographer for a while. Franklin drove by her house but didn't see any sign that he was there.

Later, in Atlanta, he heard more about the trial taking place in Lawrenceville, so he found it on a map. When he found out it was close, he drove there, checked into a motel, and started surveilling the courthouse looking for a good hidden sniper's perch. He read that Flynt liked to take his lunch breaks at the V&J Cafeteria, so he walked the route between it and the courthouse to look for a good sniper's nest. He found an abandoned house nearby. Once he fired, he thought the shot had been good enough to kill Flynt, and he was disappointed when he learned he had lived. He told us he even considered going into the hospital to finish him off but couldn't figure out a way to safely escape.

He described in detail how he planned his sniper crimes, how he would try to visualize every detail beforehand, just as an actor or dancer might before going onstage. He said he tried to think of and plan for every contingency that might come up, such as if a cop happened by or heard the gunshots. He always had his escape routes mapped out and knew how he was going to dispose of the weapon. When he only managed to wound a victim, he considered the mission a failure.

In many interviews, the nonverbal cues are at least as important as whatever the subject says. As Franklin stood before us, he would act out what he was describing, demonstrating his shooting techniques and martial arts skills. I'm pretty sure he thought he was impressing us, but as Ken and I glanced at each other, we had to control ourselves not to laugh at his absurd antics. I kept reminding myself he was proudly acting out his methods of killing innocent human beings.

What is always so interesting to me about these encounters, and Franklin was no exception, is how the subject may be very vague and claim not to remember certain features of his life, or even entire time spans. But when it comes to describing the crime itself, once they transport themselves back into that mindset, they can recall every fine detail. They may remember little about the victim's reaction, unless that is part of the signature aspect, but they can often recapitulate every move they made during the commission of the crime. And in every instance with Franklin, he could tell us exactly where he was, recall the car he was driving, if and when he got out, the model of weapon and caliber of ammunition he used, and where he aimed to get the most effective shot.

When I set up for him the murders of Darrell Lane and Dante Evans Brown, the two young boys in Cincinnati, and mentally placed Franklin on the railroad trestle looking down on them through his gunsight, he responded in a dispassionate and matter-of-fact manner. He could re-create the exact moment when he squeezed the trigger for each kill. Yes, he had been hoping for a mixed-race couple to kill, but when the boys came along, it was at least an opportunity to eliminate two Black people.

I was listening closely to see if I could detect even a hint of remorse about victims so young. The grief of the two families must have been unimaginable. And I did think Franklin spoke with a slight tinge of regret about this one, but not, in my estimation, enough for him to lose sleep over.

What was most revealing to me was not even the details he recalled so much as the manner in which he related them. There was neither triumph nor remorse. If he talked about achieving only one kill with five shots, it was as if he were a professional baseball player describing how he only went one-for-five in yesterday's game. While he was forthright and passionate about his racism and stated plainly that he felt the survival of the white race was at stake, his description of the crimes themselves was completely procedural.

When we brought up the spontaneous Manning and Schwenn murders in Madison, he shifted to telling us about how he had planned to kill Judge Simonson. But as we directed the conversation back to the East Towne parking lot, he admitted that his temper got the better of him and he was diverted from his original goal. He conceded that this had been a stupid move on his part and that he was lucky to get away. He kept coming back to Simonson and contrasted how carefully he had plotted that one out, even down to making sure he could positively identify him and not kill the wrong individual. Most violent offenders are a mass of contradictions, and these observations helped us delineate the organized and disorganized components of Franklin's criminal orientation. When he was able to keep his highly systematized belief structure intact, he was a well-organized offender. But when his hair-trigger temper was set off, though he would fall back on that belief structure for the targets of his rage, the organized aspects of his criminal personality went out the window. Yet despite his admission that going after the couple in the parking lot was ill-advised and risky, the way he told the story, it was as if the couple had gotten in the way and therefore bore the blame for his not being able to complete his mission. He seemed to imply that being able to rid the world of one more interracial couple was at least some accomplishment, though.

It wasn't just that he was personally offended by Blacks and whites socializing together. What he was ultimately afraid of was their mating, which was why he had reacted so strongly to the *Hustler* pictorial. "Miscegenation is the genocide of the white race," he said. From his tone and casual body language we got the impression it was one of his frequently quoted slogans. In fact, I had come to feel that the racist declarations and rants were a sort of shield for him, a way to protect himself from having to delve into the reasons for his own inadequacies.

Like many serial killers whose crimes are sexually based rather than mission oriented, Franklin told us he was always looking, always on the hunt, for a target. This helped explain the continuum of his activities, from highly planned crimes like the bombing in Chattanooga, to the East Towne parking lot shooting in Madison because a car in front of him wasn't moving quickly enough, to the default shooting of the two teenage cousins in Cincinnati after he couldn't spot a mixed-race couple to kill. The point is, in each case, he was prepared to take a life; he was ready for the opportunity to present itself.

The difference between him and a typical serial killer is that rather than return to the scene of the crime or body dump site, or taking souvenirs from the victim, all in an effort to relive the power, pleasures, and satisfaction of the crime, as soon as Franklin pulled the trigger or detonated the bomb, he was through with that crime and already thinking about the next one.

With the types of predators I've dealt with, the distinction between modus operandi—M.O.—and signature is very important. M.O. is what an offender does to accomplish the crime, such as bringing a gun and a demand note to a bank he intends to rob. Signature is what he does to satisfy himself emotionally in the crime. For example, with Dennis Rader, the BTK Strangler, it was binding, and watching his victims die slowly.

This was why Franklin's description of his methods was so revealing. I have to say, one of the aspects of his criminal career that would have thrown us off in any UNSUB analysis was that he employed both firearms and explosives, and that he targeted both African Americans and Jews. This is pretty unusual and probably led to what we call linkage blindness—the inability to connect two or more crimes and attribute them to the same individual, whether that individual is known or remains an UNSUB. Once we got to examining Franklin's case and background, it made sense. After his eye accident, he had trained himself to be a crack shot, which made him a skilled sniper. And from immersing himself in Nazi and right-wing extremist literature, he had learned how to make rudimentary but effective bombs. So, the M.O. wasn't important to him except in how effective it would be in killing his intended targets. No other serial killers I can think of quite fit this description. Generally, once they find one method that works for them, they stick with it. And, of course, if the means of killing is part of the offender's signature, that will not change at all, though it might easily become more elaborate over time.

Now, Franklin was no Theodore Kaczynski, the Unabomber, who felt a need to show his intellectual prowess by creating complex and elaborate bombs and sent them off through the mail to his carefully chosen targets. Franklin was not nearly that sophisticated. He had to rig up his bombs on the spot with dynamite and blasting caps and be there to detonate them. He was not concerned with the "art" of the murder method, only the result.

When we asked him how he chose whether to use a gun or a bomb on an intended victim, he really didn't have an answer. He said that a bomb has the potential to kill more people, which is what he hoped to do at Beth Shalom Synagogue in Chattanooga, but that didn't explain his decision to blow up Morris Amitay's house rather than simply wait in hiding to shoot him. We suspected it had to do with the fact that he had just set off a bomb, so he felt comfortable with it.

What it seemed to come down to, as Franklin explained it, was what he could get hold of at the moment, and in some places, explosive material was easier to come by than in others. In other words, unlike a Ted Kaczynski, someone with a genius IQ who took pride in his bomb-making craft, to Franklin, both the guns and the bombs were simply M.O.—means to an end. This versatility of M.O. had nothing to do with his level of organization; it merely served his signature—striking out against Blacks, Jews, and race mixing.

He was practical in other ways, too. He told us he would never rob a bank in a city or town where he was intending to commit a murder, or vice versa. He knew that most banks had video surveillance systems, and even if he disguised himself, as he routinely did, he wouldn't risk being linked by witnesses and security footage from the two events.

Unlike a Rader or even a Kaczynski, Franklin didn't fantasize about the act itself; that wasn't what aroused him. Rather, it was the mission, so it really didn't matter to him whether the tool was a gun, a bundle of dynamite sticks, or some other explosive. As long as it killed his enemies, he was satisfied and could move on.

Even that, I sensed listening to him, was not a complete explanation. As he roamed from place to place, he regularly picked up hitchhikers. But only teenage girls and young women, some of them prostitutes, most of them not. When I asked him about this, he repeated his line about wanting to protect them from danger, particularly the danger of Black predators, though the irony of his having been probably the most predatory individual out there at the time was lost on him.

Assessing him as we listened, what we realized was that in his own way, Franklin was just as much of a profiler as we were. Every time he picked up one of these women, he began conducting his own assessment protocol, asking questions to elicit the kind of information he found relevant to his "work." He made it clear that he thought of himself as a vigilante and he was looking at victimology, just as I did when I worked a case. The difference was that I was evaluating the victim after the fact, trying to figure out how much risk the situation had placed her in and what had made the offender choose her for his crime. Franklin, on the other hand, had the strategic advantage of having the unsuspecting young woman there in his car and under his control when he decided whether she was to be a salvation project or a victim.

Nor should these hitchhiking murders be confused with the East Towne Mall parking lot murders in evaluating organization versus disorganization. While each pickup was spontaneous, the methodology and criminal intent were not. He may have chosen whom he stopped for by their looks or how he was feeling at the moment, but what he was going to do once the young woman got in the car was all planned out, and he knew how he would qualify her on one side of his personal moral ledger or the other.

Or, to put it into a more psychosexual context, he was picking up these women and girls because they appealed to him, and then, like a stern but loving daddy or a dominant master, he got to decide whether they were deserving of reward or punishment—in this case, life or death. And here is where he hearkened back to my general experience with violent serial predators. As we have noted, he and his siblings were subjected to harsh physical and psychological punishments, often when they didn't perceive they'd done anything wrong. With these girls and young women, he was reversing roles with his parents and acting out his own childhood trauma, but on a far higher and deadlier order of magnitude. His frame of reference for the ultimate misbehavior was race mixing. So, if he found out that one of the women had engaged in it, it was his privilege and obligation to punish her, whether or not she perceived she had done anything wrong. And for almost any serial killer, even if the sexual component of the crime is secondary, having the power to punish his victim is a huge motivator and turn-on.

Though he wouldn't admit it to us, I suspected that he cruised for female hitchhikers to pick up secretly hoping they would turn out to have associated with Black men so he could mete out his form of discipline and justice.

We've found that family has tremendous influence on the development of almost all predatory criminals. What they become and do is deeply rooted in family. As we got into his formative years, it was clear that Franklin had had a pretty miserable time of it. Aside from the physical abuse from both parents, he described how his mother insisted that he and his brother Gordon come home directly from school every day and sit on the couch, watching television if they wanted to, rather than stay outside and play with other boys. Though it wasn't violent in and of itself, this was certainly a manifestation of extreme control by an unstable woman, and Franklin told us he thought it stunted his emotional growth. I wouldn't disagree with that. Frequenting prostitutes, whom you don't have to impress with charm, is a common manifestation of that stunted emotional growth in adult men. He also mentioned that his mother said her mother used to beat her as a child. Unfortunately, this is not an uncommon pattern, as we see when he started beating his second wife, Anita.

At one point, he said, he came across photographs of his mother's family back in Germany. The young boys were dressed in Hitler Youth uniforms. We could tell from his hesitation in speaking and the way his body tensed up that Franklin's conflicted views of his family were further confused and complicated by this revelation. He hated his mother and wasn't too crazy about his father, so on one level, he would have been predisposed against their families. On the other hand, knowing that he was descended from pure Aryan stock would have given him a sense of pride in his heritage and strengthened his sense of mission and purpose in fulfilling his Nazi destiny.

Franklin's memories of his childhood accident were vivid. The family was living in New Orleans at the time. Though it has been alternately reported as an incident with a BB gun and a calamitous fall off a bicycle, he told us that what actually happened was that he and Gordon were out back trying to pull the spring out of an old window shade, each from the opposite side, when suddenly it sprang out and hit him squarely in the right eye. If it had been a BB gun that caused the accident, I had theorized that his preoccupation with precise shooting might have been a psychological attempt to master and dominate the object that had hurt him so badly, but when I heard the real story, I had to abandon that theory.

As soon as he was hit, he said, he was in immediate pain and all he could see was red. He was rushed to Charity Hospital, where he spent a week recovering on the ward. When he was released, he told us, one of the doctors said to his mother to be sure to bring him back after a certain amount of time for an operation that would restore his sight. She never did, and by the time he was old enough to deal with the matter on his own, he was told the eye was too scarred over with cataracts.

He told us that sometimes in prison he dreamed of having killed his mother. When we asked if he thought his murders might have been displacement for his anger toward his mother, as Edmund Kemper's had been, he said he thought that was a real possibility.

This was one of the clues I had been looking for. There is almost always a triggering event or a series of triggering events that combine to form the catalyst that leads to criminality and violence. We had identified the Beth Shalom bombing in Chattanooga and the Amitay house bombing in Maryland as Franklin's first intentions to kill, even though each failed. Then there was the encounter in the parking lot as the event that moved Franklin from anger and hatred to the willing application of deadly force to victims right in front of him. But with the eye injury recollection, it was clear that the eye accident and its aftermath not only led to Franklin's compensatory behavior as a marksman but solidified the resentment and hatred he felt for his mother, which then needed an outlet. Had it not been for her, he seemed to be telling us, his eye could have been fixed, he would have been able to join the military or have a career in law enforcement, and he would not have had to seek his power where he did.

Of course, this is too simplistic an answer on its own and doesn't account for his rabid hatred of African American and Jewish people. But it does go a long way in addressing the nature-versus-nurture conundrum—the one we are always dealing with in our study and analysis of the criminal personality—as it applied to Joseph Paul Franklin. Given his hard wiring and the effects of his upbringing and environment, and especially his view of his mother's abuse and neglect, it was as if nature had loaded the gun and nurture pulled the trigger.

There are a number of ways people manifest bad backgrounds and upbringings. When I first started doing this kind of research back in the 1970s, we were perplexed by the fact that nearly all serial killers were men. Yes, men are loaded with testosterone, which is thought to fuel aggression, and men tend to be stronger and have better physical equipment for sexual assault than women. But when we thought about the fact that girls probably suffer from bad family situations just as frequently as boys do, we wondered if it could be that women are able to suppress their anger, rage, and resentment better than men, or are they simply not as harmed by the same kind of treatment? That didn't seem possible.

As described earlier, we found the answer is that girls are just as adversely affected, if not more so. But unlike the boys, as they grow into women, they tend to internalize their abuse and take it out more on themselves than on others. This can manifest itself in self-abusive or harmful behavior, prostitution, drug addiction, or hooking up with men who continue to treat them badly because they have been brought up to feel that they don't deserve any better.

Franklin followed the male behavioral pattern. He externalized rather than internalized his anger, first by seeking out people and groups with beliefs like his, then by going out on his own and defining himself as a man of action. It is significant to note that while all four of the Vaughn children suffered at the hands of their parents, the two boys were in continuous trouble as adults and the two girls grew up to lead relatively normal lives, largely free of the hatred and prejudice that consumed their brother.

We are always looking to connect the developmental dots, and Franklin confirmed some of these conclusions with his next recollection. As with many of the predators we've studied, Franklin had a conflicted view of law enforcement. His heroes growing up, he said, were cowboys and outlaws like Jesse

James, men who were strong and brave and acted on their own. Even into his adult years, he said, he frequently wore a western hat. But he liked the power and the heroism that a badge, uniform, and gun represented, and as a teen, he wanted to become a police officer when he grew up, as his uncle was. When he was seventeen, he said, his mom was talking to a local policeman and said that her son wanted to be a cop. The officer told her that unfortunately, that wouldn't be possible, because someone who was legally blind in one eye couldn't qualify. Franklin said that was the end of his aspiration. Coincidentally or not, it was shortly after this that he dropped out of high school, married sixteen-year-old Bobbie Louise Dorman, and joined the American Nazi Party. He had read *Mein Kampf* two years earlier.

This was a young man who was filled with anger, rage, and hate, and he had to find an outlet for it. Much of what determines what type of criminal someone like this will end up as is the life circumstances during the formative years. In Franklin's case, the combination of his inadequate and abusive parents, the family's abject poverty, the racism and discrimination endemic in the Jim Crow South in which he grew up, and his exposure to Hitler and the Nazi philosophy, all came together to shape the type of adult James Clayton Vaughn Jr. evolved into.

He made it clear that he had joined the Nazis in high school because he embraced their philosophy and only joined the Ku Klux Klan when he was out on his own. But he repeated to us that he didn't stay in the Klan very long because he was convinced it was infiltrated with FBI informers. Many of the rank and file he perceived as simply a bunch of drunks like his father.

I explained to Franklin that sexual predators often use

violent pornography to fuel their desires and motivate their crimes. Did he have anything similar that drove him? Yes, it was the white supremacist newspapers that detailed the violent crimes Blacks committed against whites. He got angry every time he read one and determined to do something about it.

Could intervention in his mid-teens have turned him around and changed the outcome? It's possible, if he could have been shown a positive alternative to Nazism and racial hatred, as well as a discernible way out of his poverty and personal desperation. Unfortunately, that's a tall order. It would have meant, at the very least, being able to get him out of his home environment, and if not out of the racist-leaning South, at least under the influence of an older authority figure or mentor who could have shown him an alternative and exposed him to African Americans on a meaningful personal level. Certainly, the majority of men and women who grew up in this era and region were able to get past the racism they had seen all around them, but Franklin was so damaged on so many levels that this was not something he could accomplish on his own through the normal course of maturing into adulthood. His racism was something he had to hang on to because he had little else to give him a sense of identity and purpose, as well as to have something to blame for his lack of success. The point is, most men who grow up in abusive, impoverished, and hostile environments don't become criminals. But we seldom see a serial killer or repeat violent offender who comes from what we would call a normal, healthy background.

Like many serial killers, Franklin looked up to other criminals who had come before him. He confirmed that he had gotten the idea of fomenting a race war from Charles Manson.

When I mentioned that I had interviewed Manson in San Quentin, his interest perked up. He wanted to know what Manson was like in person. I said that I was surprised by how short he was, but he had obviously developed the ability to capture attention with his expressions, verbal skills, and body language, just as Franklin had compensated for his crippling eye injury with relentless shooting practice.

Franklin was fascinated as I related how Manson climbed up and sat on the back of a chair so that he could physically dominate Bob Ressler and me, and even convinced Bob to give him his sunglasses, so he could tell the other inmates he had successfully conned an FBI agent. You would think Franklin didn't have much in common with a counterculture denizen like Manson. But Franklin admitted that he idolized Manson, particularly his ability to get followers to do what he wanted them to, something Franklin knew he was incapable of. Getting others to go out and kill at your suggestion was the ultimate power to him, but Franklin knew the only hope he had of influencing followers was by his example in violent action. The difference, he said, as if he had studied the subject carefully, was that Manson's strategy was to kill a bunch of rich white people and blame it on the Blacks, while Franklin was more direct—he would just kill as many Black people as he could. After we interviewed Manson, I became convinced that if he had fulfilled his ambition to be a rock star, the Tate and LaBianca murders never would have taken place. I was never even convinced he was interested in the race war he preached to his "family." It was merely a convenient cause to get them focused on. And there was no way Manson would ever embark on a career as a lone killer; it just wasn't part of his makeup. For Manson, it was all about recognition, living through other people's efforts and his power over others.

Franklin, though, was serious about a race war. He was hoping other white supremacists would see what he was doing, even though they didn't know who it was, and copy his actions. In fact, the thing that most seemed to annoy him was that he wasn't as famous or notorious as other serial killers and assassins, whom he didn't consider nearly as accomplished as him. Despite all of the media reporting, he felt the press didn't really appreciate the significance of his mission. Many sexually motivated serial killers compare themselves to others by way of both fame and number of kills. As BTK, Dennis Rader didn't think he was getting his due from the media and envied the attention directed at the Son of Sam. Franklin was neither boastful nor even particularly reflective on this point. It was as though it should be obvious to us that this was what he was all about. And though his thinking about anything involving himself, as opposed to his mission, tended to be erratic, this was clearly one of the motivations for his round of confessions now that he felt he had nothing to lose.

He also was disappointed that, as in his early days in the Klan, most of the racists turned out to be all talk and no action. When he mentioned this, it was one of the psychological factors we were most focused on. For both the Bureau and the Secret Service, the question of when do you get concerned about hateful thoughts having the potential to transform into violence is a paramount consideration. Unlike most serial killers, who are generally fantasizing about their sexually based crimes for years before they actually commit one, it was a terrifying realization that with someone like Franklin, he could always be just one conversation or hate pamphlet away from activation. Everything he complained about or implied confirmed my original and ongoing suppositions about that transition in Franklin.

The violent tendencies came from a combination of his hardwired personality traits and whatever was going on in his brain that controlled executive function. The hostility, frustration, anger, and resentment at his own failings were what pushed him over the edge into violence. As far as I was concerned from my research and study of his case, the targets—African Americans and Jews—were the justification for his violent impulses and need to act out. Terrorists may be very sincere in their sense of mission, but I have yet to see or study one that didn't have deep psychological inadequacies and a need to prove their worth that actually drove them to action. It is what terrorist leaders and strategists have learned to recognize and profile in recruiting suicide bombers, gunmen, and hijackers. That is to say, had everything else been the same in Franklin's background except for his exposure to Nazi and racist sentiments, I think he still would have developed into a killer, just with a different set of targets. Violence for someone like Franklin becomes the ultimate fulfillment, the ultimate self-actualization.

The main difference I noted between Franklin and other terrorists and assassin types was that he was not suicidal or harboring a martyr complex. All of his risk-taking was calculated. A key component of his scrupulous planning was his exit, or escape, plan for each crime, something most assassins do not think much about. Their historic act is often the endgame, and while it may end their lives or their freedom, they believe it will secure their place in history. Franklin had no such intentions. So when I brought up his initial arrest and escape in Florence, Kentucky, and whether he had any contingency plans or money, clothing, and other resources stashed somewhere, he said no; he hadn't been planning to commit a crime that night so he hadn't been prepared to have to get away.

This is a perfect example of the confusion and mixed organized and disorganized presentations we often see from repeat offenders. They think they are rational and planning their crimes, but when they are not in that mindset, they become vulnerable to their own emotions and impulses, just as the rest of us do. When Franklin saw that his car was blocked in at the motel, he became both angry and worried that he would not be able to get out when he had to. At the moment when he decided to complain about it, he said he didn't even think about the fact that there were guns in the car, because it didn't occur to him that anyone was going to look there.

The escape from the Florence police station was fairly skillful and certainly opportunistic, just as his brief escape in the Metropolitan Hall of Justice in Utah had been. But neither was planned, and Franklin's improvising skills only went so far.

This led to our asking about his time on the run. Clearly, he needed money, and I wanted to test my theory about why he didn't attempt a bank robbery, a skill in which he could feel confidence. Here he confirmed all of my assumptions. He knew he was a highly wanted man and that police departments all over the country would be on the lookout for him. Banks had surveillance cameras, and even in disguise, if he robbed a bank, it would be too easy to get to him before he could get away. He had also had a bad experience with an exploding dye pack of currency in one bank robbery and didn't want to take the chance on that.

Did he think about any other kind of robbery or theft? Vaguely, he said, but it was clear to us that his inherent paranoia had reached an acute level by then, and he admitted that he really wasn't thinking clearly. He wanted to get home to Alabama or somewhere in the South and he said he was getting kind of desperate. He had sold his blood before and knew he could get some quick cash that way.

Didn't he think he would be vulnerable at a blood bank, even if he gave a false name? He looked at us as if we really didn't understand. Sure, it made sense sitting here in prison and thinking about it all. But when you're on the run and every stare from a stranger could be the one that tips off the cops, you're not thinking clearly. He was close to the end of his rope. And though he didn't use the term with us, he indicated that he was quickly decompensating. He needed a little money to regroup, find a better place to stay than a shelter, and figure out his next move. He was extremely mobile, but he needed some money to travel.

As he had in his previous interviews with law enforcement, he described the details of hateful beliefs, and it was clear that we weren't going to change his thinking any more than he was going to change ours. This paranoid, delusional system he subscribed to, in which Jews nefariously controlled all the mechanisms of government and commerce, and Blacks were their ignorant pawns and somewhat less than fully human, appeared logical to him within his own value structure and he was not about to break out of it. With someone like this, the more you try to convince them of the actual illogic of their thinking, the more attenuated and disjointed their argument may become, but they will hold fast to their ideas, because to change their thinking would mean having to acknowledge their own inadequacy. In fact, Franklin likened his three-year run of murders before his arrest to Christ's three-year ministry before his arrest and crucifixion, a comparison he would make to others. He firmly believed he was doing the will of God and justified his goal by saying that if the Lord had wanted all peoples to mix, he would have created only one race.

The only time we really broke through Franklin's protective shell was when we brought up his daughter, Lori. She was born during his killing-spree years and would by then be around ten. We knew he hadn't seen her since he'd been behind bars, and he said his ex-wife had kept her from communicating with him. He actually grew despondent when he talked about her, something that hadn't happened when he described all the people he had killed, including the two young teens.

In this respect I found him typical of serial killers. The victims he killed were impersonal props as far as he was concerned, subject to complete objectification. His own family members were actual people, subject to real emotion on his part. Though he had left Lori's mother when the baby's birth was imminent, Franklin had now gotten older and perhaps more emotionally vulnerable. The child was no longer an abstract baby but an actual live ten-year-old girl, and his sadness about having no part in her life was palpable. He seemed to hold on to the fantasy that had he been there for her, he would have been a good and supportive father, unlike how his own parents had treated him.

Significantly, in his case, this natural human feeling was

just another manifestation of his narcissism and lack of empathy. It wasn't really about Lori missing out on having a dad. It was about Franklin not being given the opportunity to be a father, as was his right. It was all about him, rather than this girl growing up with the stigma of having a racist serial killer for a father.

We had a camera with us, as we often did for the prison interviews, and he asked us to take some photos of him and send them to her. He then positioned himself in the center of the room and put himself through a series of martial arts and bodybuilder poses. At first, I thought he was hamming it up, but he seemed completely serious. This was the way he wanted his preteen daughter to perceive him. For the first and only time, I almost felt sorry for him.

CHAPTER 18

On the way back to the motel, we stopped off at a liquor store for a bottle of rum and a twelve-pack of Coke. When we got to my room, we filled up the ice bucket from the hall machine, cracked open the bottle and the cans, and set to work filling out the two interview assessment forms. I never took notes during the interviews because it would distract the subject and take away from the critical intense connection between us. So, as soon as our interviewing work for the day was over, we would try to download from our brains everything we remembered onto paper. I had gotten to the point where when I heard a significant response from an inmate, I could compose in my head what I was going to write down on the assessment form. In this case, we talked out each entry and then Ken filled them out.

First, we tackled Trapnell.

Much more so than Franklin, Trapnell had wanted to play with us, only agreeing to speculate about his crimes in the third person, as if he were discussing someone else. This reminded me of the way Ted Bundy had interacted with investigators.

Trapnell bragged to us about how he could feign any condition in the *DSM* well enough to fool a forensic psychiatrist, and when we said, "Several psychiatrists have testified in court that you were crazy," he grinned and replied, "Who am I to refute the words of such distinguished professionals?"

We sparred over whether I'd be able to catch him if he were a fugitive. I told him I knew he'd break off any contact with his family because the feds would be looking at them, as they did with Franklin. But I said I knew his dad had been a high-ranking naval officer who he loved and respected and wanted to emulate. Gary's crime spree began when his father passed away. I told him I would have agents staking out the grave at Arlington National Cemetery at Christmastime, his father's birthday, and the anniversary of his death.

I could see his facial expression change, and in spite of himself, he broke out into an ironic smile. "You got me!" he conceded.

Then I brought up the "Free Angela Davis" demand. Smiling again, he told us that he was committed to Black liberation and righting the wrongs that had been perpetrated against African Americans since they were first brought over as slaves. But it all seems a little too glib and rehearsed. Finally, as we kept pressing him, he got down to the real motivation, just as David Berkowitz had when he eventually admitted the neighbor's dog ordering him to kill was a total fabrication.

Trapnell said something to the effect of, "I knew that this airplane hijacking had a lot of risk to it and might not work out. And if it didn't, and I was arrested, I would be doing a lot of hard time. I knew the makeup of hard-core federal prisons, and I figured if the big Black brothers thought I was a political prisoner,

I'd be a lot less likely to get cut up or get my ass raped in the shower!"

So, there it was. Despite the perverse cynicism of the idea, my respect for his criminal sophistication shot up. This man was consciously covering his contingencies.

This turned out to be an important insight in training our hostage negotiators, of whom I used to be one. Any time the subject makes a demand or statement that just seems to come out of the blue, it could be highly significant. It could mean that the negotiation dynamic has shifted and you are now nearer the endgame, that he has already moved on to the next stage in his own mind and the negotiator should react accordingly. It could prevent a lot of violence.

Now, compare this thinking to Franklin, who made elaborate plans to succeed in his crimes, knew in advance the most direct route out of town, disposed of his murder weapons, traded his cars, and changed his physical appearance through wigs, disguises, and differing hair lengths. What he didn't plan for was what to do if he were caught and imprisoned, other than trying to escape. His passionate hatred of Black people precluded any accommodation with Black inmates and on his third day at Marion, he was practically killed by African American inmates who knew his repellent reputation, while Trapnell was pretty much left alone.

In fact, Trapnell had nothing but contempt for Franklin. When we told him that we had just interviewed Franklin, he was offended. "You came here to talk to me and then went and wasted your time talking to a racist idiot like that?"

As Ken and I sat in the hotel room and next turned our attention to Franklin, we agreed that we had a pretty complete idea of his character and psychological makeup. Pretty much all of the important elements I'd speculated about in the fugitive assessment turned out to be accurate, and his description of his weeks on the run confirmed that upping his stress level forced him into a vulnerable position. His course in life was, indeed, set by a combination of the basic physiological makeup he was born with, his dysfunctional, abusive, and neglectful upbringing, the sociological customs and prejudices of his environment, and his gravitation toward extremist groups that he sensed would help compensate for his own inadequacies, bolstered by a steady diet of hate-filled propaganda literature.

But what we also concluded—Ken from his perspective in protection and trying to anticipate danger, and me from my experience using behavioral clues to solve crimes and identify suspects—was that someone like this is difficult to catch and even more difficult to predict.

As Ken put it, when you're looking at extremist organizations that either advocate or support violence, how do you determine which of this bunch of yahoos is actually going to become violent? How do you pick out those particular individuals? There doesn't seem to be any exact pattern of behavior to look for.

Franklin was correct that when he was a member of the Klan and the Nazi party, those organizations were riddled with informants. But as far as I could determine, none of the FBI's informants ever reported Franklin or suggested he would go out on his own and start killing people. To my knowledge, his name never came up in any informant report.

Attempting to change the thinking of an adult like Franklin, we concluded, was essentially an impossible task. First of all, you would have to come up with something to replace it with, and that would take a total immersion effort, like the tactics the military employs in basic training to break a recruit down and then build him or her up again with the desired attitude and outlook. That would not be possible with a Franklin. Not only would he be unwilling, but there were no facilities or mechanisms with which to accomplish it. Franklin's hatreds were what kept him going, what gave him his direction in life. Absent the idea of intensive early intervention we mentioned earlier, there was no way we were going to transform him as an adult into a useful, productive member of society with a positive outlook and sense of himself.

The fact that this was difficult to do, which we determined from our assassin study in general and our interview with Franklin in particular, is not a capitulation to the problem nor a failure of behavioral science. It is merely an acceptance of the research conclusions. From a Secret Service protection perspective, if you can't always anticipate where the threat is going to come from, then you have to harden the target. For example, in the 1940s, '50s, and '60s, it was common for presidents and other dignitaries to ride through crowds lining streets and buildings in open cars. After the horrific tragedy of John F. Kennedy's assassination in Dallas on November 22, 1963, it was instantly determined that it was not possible to protect a president that way, and the practice halted. It is a shame when a political figure has to be separated from the public, but that is the reality we must face.

This doesn't mean the Secret Service should stop keeping track of threats. It just means that additional layers of security are essential.

My needs and agenda as an FBI agent were somewhat different. While I would love to be able to root out potential killers before they strike, that is not our main function. The FBI is the Federal Bureau of Investigation, not the Federal Bureau of Prevention. We always say that if you are relying on us in law enforcement to solve your societal problems, you are already too late. Our behavioral science efforts are aimed at learning more about how to catch them once they have acted and, hopefully, before they act again. And from that point of view, the interview with Franklin was useful and illuminating.

Joseph Paul Franklin defined a different type of serial killer and assassin-type personality for us. He came from a similar—if not considerably worse—type of background than other offenders we'd studied, but his crimes were not primarily sexually motivated. Though he was certainly compensating for his own inadequacies, he was not primarily seeking his own glory and place in history so much as trying to alter society to match his own vision, through his own efforts and his example to others. He took calculated risks, but he did not want to be caught or have a martyr complex. And though his ideas were abhorrent, he sincerely believed what he was doing was religiously motivated. He genuinely believed that the combination of his planning and luck was the work of a God who wanted him to succeed in his mission. He moved freely around the country using a variety of aliases, fake IDs, disguises, different cars, different weapons, and different addresses. He was not limited to a single means of killing. He committed most of his crimes from a distance, so there was little physical or behavioral evidence at the point of the murders and therefore the crime scene had to be considerably expanded. For all of these reasons, someone like this is very difficult to catch until he makes a mistake. And just as difficult to determine is who is going to emerge from the crowd as the next Joseph Paul Franklin.

Fortunately, there haven't been many killers as single-minded, versatile, mobile, and resourceful as Franklin. But perhaps the most important takeaway from the interview was to expand our investigative horizons in terms of linkage and pattern recognition; we needed to coordinate more fully with law enforcement agencies in the areas of the crimes and share and compare evidence and reports; and when an UNSUB of this type was operating, we needed to come up with proactive strategies to get him to make a move that might identify him, and place him under as much pressure and stress as possible to force his hand.

We weren't the only ones who continued to be interested in Franklin.

WHILE FRANKLIN HAD LARGELY STOPPED CONFESSING TO CRIMES IN THE MID-1980s, there were still unsolved murders that had been circulating around him for years—crimes for which he was the only suspect. And one in particular would prove crucial to his fate.

Almost as soon as Franklin arrived at Marion in 1982, Lee Lankford, a captain with the Richmond Heights Police Department in St. Louis, started writing to him, asking to talk to him about the Brith Sholom synagogue shooting back in 1977. When Franklin was initially arraigned in Salt Lake City, Lankford had followed the case, and the details led him to contact the Justice Department in Washington to see if they would share their files with him. As he reviewed those files, everything pointed to

Franklin as the synagogue shooter, but the hard evidence still wasn't there. What Lankford really needed was a confession.

So, he started reaching out to Franklin directly, at first sending him small amounts of money for the prison commissary and magazines to relieve his boredom, trying to build some kind of rapport. At one point, he drove the 125 miles to Marion and showed up at the penitentiary, asking to see the inmate, but Franklin refused.

But Lankford never gave up, working the case for years. He spoke to Gerald Gordon's mother every week. When he was appointed chief of the department in 1988, he moved a large box filled with files on the case into his new office, continually trying to give prosecutors what they said they needed to bring a charge in the Gordon murder. He never stopped trying to get to see Franklin. "I want to sit down eye-to-eye with the guy," he told the *St. Louis Post-Dispatch*.

In October 1994, Franklin contacted the FBI to say he was the shooter in the murder at Brith Sholom Kneseth Israel Synagogue almost exactly seventeen years before. The Bureau passed along the information to Richmond Heights PD. Lankford had retired as police chief two years before, but no one was more gratified than he that they might finally have closure and he could fulfill the promise he'd made to Gerald Gordon's mother so many years before. He had even made a trip to Irving, Texas, to visit the gun shop where Franklin said he had bought the rifle. Franklin had robbed a bank in Oklahoma in which the stash he escaped with had exploded with a blue dye pack. Franklin put two of the hundred-dollar bills inside his socks, figuring his foot perspiration would fade the dye. The gun shop owner told Lankford that Franklin had paid for the rifle with two creased and soggy hundred-dollar bills that he took out of his shoes.

Detective sergeants Richard Zwiefel and John Wren went to Marion to interview Franklin the following month. "He said he wanted to clear his conscience. It was a calm, almost casual conversation," Zwiefel said. "He wasn't cocky. And he is still a racist."

While on the one hand it was surprising that Franklin would be confessing now, he seemed to have a reason. Back in 1983 and 1984, when Franklin contacted authorities in Georgia, Tennessee, and Wisconsin, he had never mentioned this crime, even though investigators had considered him a strong suspect. One possible reason was that Missouri was a death penalty state and, as he put it, he "wasn't anxious to be sentenced to the gas chamber."

So, what changed in those ten years?

Well, for one thing, he said he had a dream in which he was told to confess, and Franklin, professing a great faith in the Bible, was a strong believer in signs and portents. On a more practical level, though he was still housed in the protective K Unit at Marion, Franklin remained convinced that both the Black inmates and the guards wanted to kill him. Even if he was convicted and sentenced to execution, he figured he would live a lot longer on death row in a state prison as the wheels of justice ground slowly than he would if he remained a target at Marion.

And there was a third factor, which may have played a role: 1994 was the year that Franklin's father, James Clayton Vaughn Sr., died in a mental ward in Biloxi, Mississippi.

After the indictment, Franklin was transferred from Marion to the St. Louis County Jail. At the time of his confession, the Richmond Heights case was only his third known murder chronologically, after the killings of Alphonce Manning Jr. and Toni Schwenn in the East Towne Mall parking lot two months earlier, but in many ways it was his most archetypal crime—a well-planned sniper attack from far away with no particular target in mind; just a general vicious hatred of an entire group of people. It is as if it was also the most important act of fatal violence to Franklin himself, representative of his outlook on life and sense of his own purpose. Sometimes, despite law enforcement's best efforts, it's not enough just to know who the killer is to bring him to justice for a particular crime.

But in the end, I suspect, the Brith Sholom Kneseth Israel murder was so central to Franklin's psyche that even though Missouri was a death penalty state, he couldn't keep from claiming it as his own; the compulsion was that strong. All the other reasons he gave from time to time for the confession, including wanting to get out of Marion, I think, were side issues.

IT WOULD BE MORE THAN TWO MORE YEARS BEFORE HE CAME TO TRIAL IN ST. Louis. In the meantime, on April 20, 1995, Chattanooga police detective Tim Carroll took a phone call from a caseworker at the St. Louis County Jail who said an inmate there wanted to confess to an old murder.

The night of July 29, 1978, twenty-year-old William Bryant Tatum was shot and killed in the parking lot of a Pizza Hut. His eighteen-year-old girlfriend, Nancy Diane Hilton, was wounded. Tatum, a junior at the University of Tennessee at Chattanooga who went by his middle name, was Black. Hilton, who worked at the Pizza Hut, was white. They planned to get married in November. Carroll remembered the case well. It was Chattanooga's only unsolved murder that year.

Franklin said he was responsible and asked Carroll to come meet with him. Carroll asked him why he decided to confess now. Franklin said he wanted to have the most death penalty cases pending against him of anyone. The response dovetailed with what I'd learned from my interview: that he was still hungry for recognition and still compared his image to that of other well-known serial killers.

Carroll agreed to come to the jail and five days later arrived with Detective Mike Mathis. When Franklin was brought into the conference room, his head shaved, he saw Mathis and demanded that he leave the room. Franklin said he had never spoken with Mathis before and therefore didn't trust him. In the interest of getting a confession, Mathis agreed to leave.

"I was on a search and destroy mission for race mixers," Franklin told Carroll as he described spotting and following the couple. He parked his car and found a stand of tall grass nearby. It was close to Hilton's 1974 Ford Mustang. Ten minutes later the couple emerged from the restaurant. Franklin aimed and fired his twelve-gauge pump-action shotgun. He hit Tatum in the chest and the blast tore through his heart and one lung. Hilton was struck on her right side.

"Had he not called, the case would have probably gone unsolved," the AP quoted Carroll. "His call was to let us know he did it, not to say, 'Look, I'm sorry I did it.' There was no remorse."

A grand jury indicted Franklin for murder on March 1, 1996. When a Hamilton County, Tennessee, Criminal Court judge set a trial date the next year, Franklin commented, "That's cool."

Prosecutors said they would seek another death penalty as insurance.

CHAPTER 19

While waiting to stand trial in Missouri, Franklin began talking again.

He had admitted to a *St. Louis Post-Dispatch* reporter in November 1995 that he shot Vernon Jordan but wouldn't give details or say any more about it. The following April, he elaborated in an interview with *Indianapolis Star* reporter R. Joseph Gelarden, published on Sunday, April 7, 1996, saying that he had first considered killing civil rights leader Jesse Jackson in Chicago, but focused on Jordan when he found out he would be speaking in Fort Wayne.

He revealed that in the darkness, he wasn't sure whether the man in his gunsight was Jordan or not. But since it was a Black man with a white woman, this looked like a prime target.

Franklin also owned up to the murders of the two African American men in Indianapolis in January 1980—Lawrence Reese and Leo Thomas Watkins. "I like Indianapolis," Franklin said. "I had a good time there. The cops didn't like what I did, but I had a good time there. The alleys in Indianapolis are a sniper's dream." He said that he decided to kill for the second time there when he didn't see any news coverage of the first killing. Once again affirming how he, like so many killers, followed the media reports of their crimes.

After the Watkins murder, both cases finally made the news. That was when he said he destroyed the rifle and scattered the pieces in different locations.

In the same Sunday edition of the *Star*, Gelarden ran a write-up of a phone interview he did with me, seeking to answer what makes someone into a serial killer. He quoted from the section of the *Crime Classification Manual* in which we had used Franklin as an example of the Political Extremist Homicide category, and then elaborated on some of my ideas about the effects of abusive or neglectful upbringings on certain vulnerable personality types.

"Franklin decided that as an adult he would not be deprived of the need to belong and to be special that his childhood had denied him," Gelarden quoted me. "This lack of attention and the feelings of insignificance it produced required that he take special measures to give his life value and communicate the central theme of his life: cleaning up America.

"Franklin decided his message would be taken seriously if it was spelled in blood."

When reporters contacted three jurors from the Vernon Jordan assault trial and told them of Franklin's admission, they said they were not surprised, but that the federal prosecutors had not presented sufficient evidence to convict him beyond a reasonable doubt.

In addition to these confessions, Franklin owned up to a crime in Mississippi that someone else had claimed responsibility for. At around 9:30 A.M. on Sunday, March 25, 1979, two people washing their cars at a Jackson, Mississippi, self-service car wash discovered the body of an African American man local police identified as Johnnie (sometimes spelled Johnny) Noyes, twenty-five, who had lived nearby, sprawled beside his car. He had been shot once in the chest, apparently while drying his car. Noyes was the eldest of eight children, a student at Jackson State University who had decided he wanted to pursue a medical career after college. He was recently divorced and had two young daughters.

His mother, Emma, knew something was wrong when he didn't show up for lunch at her home in Hallandale, about ninety miles away. She had prepared catfish, his favorite dish.

Her sister in Jackson called with the horrible news.

Police could not determine a motive; they couldn't find a weapon; there were no witnesses and no suspects.

The case remained open with no significant leads until 1984, when Henry Lee Lucas, a forty-seven-year-old serial killer and lowlife drifter from Blacksburg, Virginia, started confessing to hundreds of murders around the country while being held in a Georgetown, Texas, jail. This sent law enforcement agencies scurrying, seeing if they could close any of their unsolved cases.

Jackson police detective David Fondren was one of them. He traveled to Texas to interview Lucas, who claimed credit for three murders in Jackson, including Noyes. Lucas said he spotted Noyes as he was driving down Interstate 20 in Mississippi, where the highway swings close to the intersection of U.S. 80 and Valley Street, where the car wash was located. He told Fondren he shot him with a .357 Magnum handgun.

When Fondren returned with his information, the story seemed believable, and the police were motivated to solve the case. They informed Emma Noyes they thought they were close. But the more they looked into it, the more Lucas's story began to unravel. Fondren was able to confirm that Lucas was elsewhere on the day of the murder. The proverb about "honor among thieves" seems to apply here, and each one has his own sense of it. Lucas—who, like Franklin, had an alcoholic father, had severely injured an eye in a childhood accident, and had spent much of his life as a drifter—probably murdered between three and eleven victims, though he claimed over a hundred. The motivation was probably a combination of boredom and notoriety, since he had no other accomplishments in his life. Franklin, just as much of a lowlife, did have a sense of honesty about his own crimes. It is a fact of law enforcement that unsolicited and uncoerced confessions are common, and it is part of detective work to separate out legitimate from false claims. That is why it is so important to know as much as possible about the suspect to determine his overall credibility. Though Franklin certainly played the system and often denied statements he had made earlier, I never knew him to claim a crime that wasn't actually his.

That's where things stood until early May of 1996, when Franklin reached out from his Missouri jail cell and claimed credit for this one, too. Jackson homicide detective Chuck Lee investigated the claim and said Franklin offered details only the killer would know. "We're convinced and the district attorney's office is convinced he's the right guy," Lee told the Jackson *Clarion-Ledger*.

He discussed with the D.A. going to Missouri to interview

Franklin, just as Fondren had interviewed Lucas, but they concluded that there was no point in beginning another legal proceeding and decided to finally close the case and give the Noyes family the certainty they had all sought for seventeen years. Emma Noyes passed away in November 1997, finally knowing who had killed her son.

THE TRIAL FOR THE MURDER OF GERALD GORDON BEGAN ON MONDAY, JANUARY 27, 1997, almost twenty years after the crime. Douglas J. "Doug" Sidel led the prosecution. Acting as his own attorney, Franklin was assisted by public defenders Karen Kraft and Richard Scholz, but he wouldn't let them intervene or offer objections to any of Sidel's witness questions. The jury was all white and all male. Franklin maintained he excluded women because they are more compassionate than men and would be less inclined to impose the death penalty he claimed to seek.

Despite the fact that Franklin had already confessed to killing Gerald Gordon and wounding William Ash, the prosecution laid out a carefully constructed timeline showing how Franklin bought the Remington .30-06 rifle in Dallas with money he had obtained by robbing a bank in Little Rock, already intending to kill Jews with it. He considered several possible areas in the Midwest before deciding that St. Louis had the largest Jewish population. He found Brith Sholom Kneseth Israel Congregation in the phone book, along with several other synagogues, drove around to look at them, and chose this one because of its easy access to the highway for his getaway.

William Ash, who had lost a finger from a gunshot, and Steven Goldman, who had been near Gerald Gordon in the synagogue parking lot and had been grazed by another bullet, testified. It was the first time either one had seen Franklin in person.

One of those sitting in the courtroom each day was Richard Kalina, now a married, thirty-two-year-old father of two. It was Kalina's bar mitzvah Gerry Gordon had been leaving the day he was murdered. Kalina went through therapy in his twenties to deal with the post-traumatic stress he suffered from the day he said he lost his innocence and his "whole life changed in a matter of seconds."

The prosecution screened a taped interview for the jurors in which Franklin, referring to the Jews, declared, "I believe they are responsible for all of the evil, either directly or indirectly, in the world."

When Sergeant Zwiefel asked if he was sorry for any of his actions, he replied, "I'm just sorry it's not legal."

"What's not legal?"

"Killing Jews."

Franklin, wearing an orange prison jumpsuit and ankle shackles over his high-top black sneakers, cross-examined the state's witnesses, but called none of his own and didn't even present a defense.

"It was a premeditated, deliberate killing, and he wished he had killed more people," Sidel stated in his summation. Franklin gave neither an opening nor closing statement.

On Thursday, January 30, the jury took less than forty minutes to return a verdict of guilty to the charge of first-degree murder in the death of Gerald Gordon and two counts of first-degree assault. When the verdict was read, Franklin gave the jury a thumbs-up sign and mouthed, "Right on!"

The next day, jurors came back to the courtroom to decide

on the sentence. Sidel addressed them first, saying, "This act was absolutely sick and depraved. And because of it, there are three young girls who grew up without their father, who will never have their father walk them down the aisle at their weddings. Death is the only appropriate response."

When it was Franklin's turn, he looked at Sidel and said, "What he just stated was true. Shortly after I arrived here at the St. Louis County Jail, I had a conversation with an inmate up here who told me, 'You know, Franklin, if they don't give you the death penalty here, you should kill somebody to make sure they do the next time.'

"And I thought about that statement, you know, and I decided that if you guys did not vote for the death penalty, then that's what's going to happen. I'm already doing six consecutive life sentences already, plus some more time. And it would just be a farce if you guys did not sentence me to death. So that's all I'm asking. I don't have a lot of argument about that. Thank you. That's all."

They deliberated for a little over an hour, deciding that the crime fit the legal parameters for capital punishment. They also recommended two more life sentences for shooting at the two other men. The death sentence would now take precedence over the six consecutive life sentences Franklin was already serving for the Madison and Salt Lake City murders.

In opting for the death sentence, according to the court record, "the jury found the existence of three statutory aggravating circumstances: 1) that Franklin had a substantial history of serious assaultive convictions; 2) that Franklin, by his act of murder, knowingly created a great risk of death to more than one person in a public place by means of a weapon which would normally be hazardous to the lives of more than one person; and 3) that the murder was outrageously or wantonly vile, horrible or inhuman in that it involved depravity of mind."

On February 27, Judge Robert L. Campbell officially imposed the sentence, and on March 10, Franklin's public defenders filed a notice of appeal. On June 16, 1998, the Supreme Court of Missouri affirmed the conviction and sentence. After almost twenty years in prison, Franklin was facing the prospect of the death house, and it was due to his own voluntary confession.

Brith Sholom Kneseth Israel Congregation honored Lee Lankford for his dedicated pursuit of justice, arranging for the planting of a grove of trees in his name in Israel.

When asked by reporters for a comment, both Hal Harlowe, the district attorney who had prosecuted Franklin for the East Towne Mall murders, and Archie Simonson, the judge he had gone to Madison to kill before he got distracted, said they didn't believe in the death penalty but couldn't work up much sympathy for Franklin.

He had finally achieved one of his goals, however: after the formal sentencing on February 27, he was transferred to the Potosi Correctional Center in Mineral Point, Missouri. He would not be going back to Marion.

CHAPTER 20

More than a few of the deadly predators I've helped hunt and analyzed have been sentenced to death, and every time that happens, I've confronted my feelings about the ultimate punishment. I've never believed that execution is "legalized murder," because that equation ignores the critical distinction between the victim and the murderer, which I think demonstrates a moral vacuity. Though I don't believe the death penalty is appropriate for all, or even necessarily most murder cases, for the types of monsters I often deal with, I consider it fundamentally moral for society to state that it will not tolerate in its midst people who have done such things.

For me, Joseph Paul Franklin certainly fit into that category. He had destroyed the lives of men, women, and children to serve his evil purpose and assuage his own self-image. As far as I was concerned, he had forfeited his right to any claim on earthly mercy, and whether there was any mercy awaiting him in any afterlife, well, that was above my pay grade.

Still, whenever you're contemplating the end of another

person's life, it is, or should be, a sobering experience. I had my own ritual whenever a death sentence was pronounced on one of "my" murderers or he faced imminent execution. I would go to my office, pull out the case file, and sit at my desk. I would stare at the crime scene photographs. I would review the autopsy protocols. I would read the detectives' reports. And I would try to imagine what every victim was going through at the moment of his or her death. My last thoughts were always for the victims.

But while I was just beginning to contemplate the impact that my encounter with this killer would have on me, Franklin wasn't done revealing just how prolific he'd been or, from my perspective, glorying in his feats of devastation and ruin. Franklin may have been sentenced to death, but he was still intent on taking claim for every murder he had committed. And I could not close the psychic books on him.

After the Gordon murder trial, Franklin once again reached out to the media. In an interview for the syndicated television program *Inside Edition* in March 1997, while he was still housed in the St. Louis County Jail, Franklin said he had killed Raymond Taylor, the twenty-eight-year-old Black manager of a Burger King in Falls Church, Virginia, on August 18, 1979. He had not previously been a suspect in the seventeen-year-old case.

A month later, in April 1997, Hamilton County, Ohio, assistant prosecuting attorney Melissa Powers came to interview Franklin about the killing of the two young African American boys, Darrell Lane and Dante Evans Brown, in 1980. DA Powers did a masterful job of staging the interview with Franklin, making him feel in control, even as she worked him toward a confession. But the way she got the conversation going to begin with was by explaining to him that at the time of the Lane and Brown murders in 1980, Ohio was between death penalty statutes, so he wouldn't face execution in the state if convicted of the murder of the two Cincinnati teens. That turned out to be crucial. He then leaned in toward her across the table and said, "You know I did it. I killed those two dudes."

He told Powers how frustrated he was that no one was taking notice of his overall plan, which caused him to escalate. "I was just getting very upset because the news media, the national media, wasn't covering what I was doing. I guess they were afraid they'd start a race war or something. And that's what I was trying to do."

He later explained the reason this was one crime he had never wanted to confess, but that what Powers told him about the death penalty opened him up. "I didn't want to be on death row in Ohio," he told Kristen DelGuzzi of the *Cincinnati Enquirer* during a phone interview. "They have the electric chair there. I don't want to die in the electric chair."

While speaking with Powers, Franklin also confessed to the June 15, 1980, murder of Arthur Smothers and Kathleen Mikula as they walked across the Washington Street Bridge in Johnstown, Pennsylvania. Powers found him much more animated and "turned on" about this crime and speculated that in addition to his self-styled mission, the murder of mixed-race couples had an additional sexual component for him. There is often a secondary motivating factor in violent crime because violent offenders tend to equate power with sexual potency, as we saw in the case of David Berkowitz. Then there is the basic fact, which Franklin admitted to, that being in prison he was deprived of any kind of female interaction, so when an attractive woman was willing to visit him, he certainly took advantage of the opportunity, thereby allowing Powers to use her looks and intelligent charm to her own investigative advantage.

On Tuesday, April 15, 1997, Powers and her fellow DA Joseph Deters held a news conference in which they announced their intention to present evidence and seek an indictment from the grand jury so they could bring him back to Cincinnati to stand trial for the murders of Lane and Brown. Deters had been haunted by the murders. He grew up less than a mile from the railroad trestle from which the killer fired and was in law school when the crime took place. And it was lost on none of the officers and detectives who had worked so doggedly on the case that the time elapsed between the murders and the confession was three years longer than the older of the two boys had lived. Deters, who already had a solid reputation as a crusading prosecutor, said he owed a trial to the community and the families. "These two kids were citizens of this county when this derelict came into our community and snuffed out their lives," he said angrily. "Their families are still grieving. They still need and deserve justice."

Powers related the information about the Johnstown murders to Cambria County, Pennsylvania, officials, after which Assistant District Attorney Kelly Callihan and Johnstown police detective Jeannine Gaydos contacted him by phone and carried on several conversations over three months, establishing rapport and a bond of trust. Subsequently, Franklin agreed to meet with Callihan and Gaydos. During their interview with him, during which he was shackled with eight prison guards nearby, he agreed to tape a confession to the Smothers and

Mikula murders. Assistant DA David Tulowitzki and Chief of Police Robert Huntley followed up and were convinced that Franklin was telling the truth. Tulowitzki reported that Franklin sounded increasingly excited as he described the murders.

"As he explained the murder," Callihan told Tom Gibb of the *Pittsburgh Post-Gazette*, "you could see this excitement in him as he relived it." She speculated publicly that his motive might simply have been a transfer to a jail where he could attempt another escape.

Ultimately, Cambria County decided not to try the case due to the expense, the risks, and the logistical challenges of moving Franklin again, his history of escape, and their inability to add meaningfully to the life and death sentences he had already racked up.

NOT ALL OF FRANKLIN'S MURDER CLAIMS PANNED OUT OR COULD BE CORRElated with evidence. Authorities started speculating that since he knew he was behind bars for the rest of his life, he was trying to get his kill score up to record levels.

In July 1997, authorities in Nashville, Tennessee, said they couldn't link Franklin to the death of a woman named Mary Jo Corn, whose decomposed body was discovered by hunters in the Harpeth River on August 28, 1977, despite his claim to a *Post-Dispatch* reporter of having been responsible. Franklin had said he had "picked up a white woman at a truck stop along Interstate 65 near Nashville." He said he decided to kill her after she said she had slept with a Black trucker. He took her to a "wooded area near a creek, pushed her in and shot her" in the head with a .41-caliber Smith and Wesson handgun, which is a pretty rare weapon. But Corn had been strangled. It was possible

Franklin was talking about a different victim whose body had never been found or was in a different jurisdiction, and Metropolitan Nashville Police detective Brad Putnam said he had sent out a teletype message asking other departments if they had any other unsolved murders of white females from 1977.

A few days later, investigators in Robertson County, just north of Nashville, concluded that Franklin's description appeared to match the unsolved murder of Deborah R. Graham, whose bound, gagged, and bullet-ridden body was found by fishermen floating in Sulfur Fork Creek on November 17, 1977.

"Based on the basic information reported in the newspaper, it seems to match up," said county sheriff's captain Bill Holt, who had worked the case as a Tennessee Bureau of Investigation agent when the body was first discovered. He confirmed that Graham had been shot with a .41-caliber firearm. He had previously helped clear the prime suspect in the case, a trucker who had been seen with Graham at a truck stop on Interstate 65 several days before she died. But that left law enforcement with an unsolved murder.

By the end of the month, though, Franklin had clammed up. "He doesn't want to be interviewed at this time," Bill Holt announced. "We're just at a stopping point right now." It was rumored Franklin was dissatisfied with his treatment at Potosi.

Next on Franklin's list was DeKalb County, Georgia. He told the district attorney's office that he had killed Doraville Taco Bell manager Harold McIver on July 12, 1979, and fifteen-year-old prostitute Mercedes Lynn Masters on December 5 of the same year. McIver had been enjoying the first day he'd taken off work in three weeks when he came in on a Sunday evening to fix a problem with one of the cash registers. He was shot twice as he left the store, once in the left side of his chest and once under the right shoulder blade. Police quickly ruled out robbery; he was shot from a distance and he was carrying cash when they arrived. Police began to consider Franklin a suspect around 1981, when his other sniper attacks became known. But as with many of his crimes, there just wasn't enough evidence to proceed.

Mercedes Masters's body was found on Christmas Day, 1979, near Lithonia, east of Atlanta. She was lying in the backyard of an abandoned house and discovered by prospective land buyers looking at the property. She had been missing three days when her mother heard reports of a pair of boots found near the body, marked on the insides with the initials MM. She had a gunshot wound in the back of her head. Police had no motive for the murder.

Up to his old tricks, Franklin said he was only willing to confess to these crimes in person to "an attractive white female investigator," so in March 1998 DeKalb District Attorney J. Tom Morgan assigned Assistant DA Carol Ellis to go see Franklin and asked police lieutenant Pam Pendergrass to accompany her. The Associated Press reported that Morgan compared Ellis's "role to that of Jodie Foster's in the movie 'The Silence of the Lambs,' in which she plays an FBI agent sent to interview the demonic killer Hannibal Lecter." Ellis was well-suited for the job. Both attractive and a crack shot, having previously served as a DeKalb police officer, she worked in the Crimes Against Children and Crimes Against Women units of the DA's office and had a reputation for being cool under pressure.

She and Pendergrass smiled at Franklin and acted impressed as he spoke. Ultimately, they were able to get a videotaped confession out of him, in which he said he spent several days with young Mercedes and learned from her that she had had sex with Black men. "She went out with gay dudes. But once she told me she [had sex with Black men], that's when I decided to kill her," Franklin told the two investigators. He said he took her out to a wooded lot and shot her in the head and back. He decided to kill McIver because he had white girls working with him and he was sure the African American man had tried to take advantage of them.

Franklin was indicted for both murders on Thursday, April 3, 1998.

Morgan called Franklin "the most evil individual I have ever come across." He said there was no point in trying Franklin because the crimes would not qualify for the death penalty under Georgia law and because he was already facing execution in Missouri. "We just wanted to solve these crimes," he said. "We've contacted the family members and they are relieved. This has been gnawing at them for nineteen years."

In late October 1999, from his cell at Potosi, Franklin told Atlanta PD homicide sergeant Keith Meadows and detective Tony Volkodav that he was the gunman in the murder of Johnny Brookshire and his wife Joy Williams in northeast Atlanta around 8:00 P.M. on February 2, 1978, when both were shot. Johnny was killed. Joy, who was pregnant, was wounded and ended up paralyzed from the waist down. Johnny, twenty-two, was Black. Joy, twenty-three, was white. The case had remained open for more than twenty years. For Franklin, it was an all-too-familiar motive; he said he didn't like seeing a Black man and white woman walking down the street together. The murder took place the month before he shot Larry Flynt in nearby Lawrenceville.

"He knew facts about the murders that only the murderer could have known," Meadows told a local television reporter. "He was able to tell us exactly what the victims were wearing at the time the shooting took place."

Again, there seemed no point in incurring the expense and complication of bringing Franklin back to Georgia for another trial that would interrupt the process of moving him closer to execution in Missouri.

AND YET, DESPITE ALL OF HIS TALKING, THERE WAS STILL ANOTHER CASE YET to settle, a double murder Franklin had first mentioned when he began his confessional acrobatics in 1984, a crime he had brought up again when he spoke with Melissa Powers in 1997, and yet again when he spoke with Carol Ellis in 1998. His assertion that he had killed two young female hitchhikers—white women who told him they'd either dated Black men or were open to the idea—and left their bodies on the side of a rural road in West Virginia only grew more insistent over time.

Even so, it took until 2000 for the case finally to be resolved, and it remained the most complex and muddled crime investigation ever associated with Franklin as well as the one that inflicted the greatest legal jeopardy on other people. So many of Franklin's confessions had been in cases where he was already a suspect, or police had had no significant leads. The murders in West Virginia were different, demonstrating how easily a murder investigation can go off the rails, as I have seen too often in my career.

On Wednesday evening, June 25, 1980, a man on his way home drove past the bodies of two women lying side by side in a small clearing near Droop Mountain in West Virginia. When he stopped to check on them, he discovered that both had been shot. Both were fully clothed and the medical examiner found no evidence of sexual assault. They couldn't be immediately identified, but under her sweatshirt, one was wearing a T-shirt with a rainbow design, which led investigators to believe both women were probably on their way to the Rainbow Family gathering, a sort of counterculture be-in expected to attract about ten thousand participants from July 1 through 7 in Monongahela National Forest, about forty miles away. Since 1972 the group had been meeting annually to spend a week getting back to nature. This was the first such gathering to be held east of the Mississippi.

Not everyone wanted the Rainbow Family in their midst. West Virginia governor Jay Rockefeller said he wished the family would stay away, and Secretary of State A. James Manchin publicly complained that group members looked "like a bunch of gypsies." Part of what people objected to was the frequent complete nudity at these gatherings, something officials didn't think would mix well with the indigenous Appalachian community. Rainbow Family members had reported several shots being fired into their preliminary camp, presumably by locals, so maybe one of them had gone too far.

West Virginia state medical examiner Dr. Irvin Sopher informed Pocahontas County prosecutor J. Steven Hunter that both victims had been shot twice in the chest and one had also been shot in the head. The weapon was likely a high-powered rifle.

The Rainbow gathering went off as planned, a throwback to the Summer of Love and the flower children of the 1960s. But the murder of the two still-unidentified women hung over the festivities.

The gathering was over by the time the two murder victims were identified, on Friday, July 11. They turned out to be nineteen-year-old Nancy Santomero of Huntington, New York, and twenty-six-year-old Vicki Durian of Wellman, Iowa. Nancy's sister Kathy and later Vicki's brother Joseph identified the bodies after Kathy saw a sketch she said resembled her sister. Kathy was actually supposed to meet up with her sister at the gathering, and they'd travel home together. When Nancy didn't show, Kathy hoped it was because she'd made other plans at the last minute. After the gathering, though, when Nancy hadn't been in touch with anyone, Kathy looked at the sketch and made the terrible trip back to West Virginia to confirm that the police had found Nancy. Kathy told police her sister and Vicki had been traveling together to Rainbow, possibly with another woman she only knew was named Liz, which raised the question: Was there a third victim out there somewhere?

With identification came the larger impact of the violent deaths. Nancy's mother, Jeanne, said her daughter was passionate about environmental causes and wanted to use her life to effect change. Vicki was a licensed practical nurse who went out of her way to help people.

A few days later, the third woman believed to have been traveling with them was identified as Liz Johndrow, nineteen, who parted from her two friends at a truck stop near Richmond, Virginia, after getting a premonition that she shouldn't go to the Rainbow gathering. The last thing Nancy and Vicki said to her before resuming their hitchhiking was "Be careful!" Liz went back home to Northford, Connecticut, a suburb of New Haven, expecting to meet up with the other two, whom she described as free spirits, after Rainbow at her father's and brother's houses in Vermont. She contacted police after she heard about her friends' deaths and learned there was fear she had been murdered, too.

In December, on the last day of deer hunting season, hunters found the two women's backpacks in underbrush near Clifftop in Fayette County, West Virginia, about eighty miles from where the bodies were discovered. This confirmed investigators' belief that the perpetrator or perpetrators must have been locals who knew the area well. It fit in with the theory that the Rainbow people were not welcome and that the murders were the next step after the warning shots that had been fired into their camp.

What followed was a twenty-year legal spectacle in which a group of people were charged and tried for the murder. The theory of the case, based on what, in retrospect, appeared to be dubious accounts by various supposed participants ratting each other out, and well-meaning but questionable eyewitness accounts, was that seven local men in various pickup trucks and a van learned about the two "hippie girls" hitchhiking to Rainbow, picked them up in the van, drove them to the entrance of Droop Mountain Battlefield Park, and demanded sex. The girls refused, threatened to go to the police, and were then driven to a clearing in the woods, forced out of the van, and shot.

I can't think of a case I've investigated that was this convoluted and involved so many offenders working together. There is a principle in philosophy and science known as Occam's razor, which states that all factors being equal, simpler hypotheses tend to be better than more complex ones. This applies to criminology as well. If you have to jump through a lot of logical hoops and assume a lot of connections, your explanation probably veers more toward conspiracy theory than truth.

Local investigators and prosecutors were aware of Joseph Paul Franklin's confessions beginning in 1984, but discounted them because they were convinced the offender had to be local, that Franklin had merely read about the crime and claimed it for his own purposes. They said his account was riddled with inconsistencies and that he got some of the details wrong when he drew a map of the area and where he said he had dumped the bodies.

In July 1992, charges against all seven local men were dropped because "improper investigative procedures" employed by the police "seriously compromised the case and were going directly to the credibility and sustainability of the evidence on which they were obtained," according to the prosecutor.

The next year, in January, five of the men were reindicted, including the supposed triggerman, Jacob "Jake" Beard. Beard's trial began on May 18, 1993. He took the stand himself and denied all of the accusations. He said he had no idea who had killed the Rainbow girls.

Beard was convicted in June and sentenced in July to two life terms without possibility of parole. Over the next year the charges against the others were dropped. Beard's attorney petitioned the West Virginia Supreme Court for a new trial, based on Franklin's confession, but was turned down.

Not everyone in West Virginia was convinced they had the right man, though. One investigator who believed Franklin to be the real culprit was Deborah E. DiFalco, a top-flight detective and the state's first female trooper. She'd met with Franklin at Marion, and after several denials, he told her he had killed the

Rainbow girls. In a report dated January 2, 1986, DiFalco wrote, "I feel that Mr. Franklin had the motive, opportunity and capability to be the perpetrator of the crimes against the victims, Nancy Santomero and Vicki Durian."

DiFalco was not the only one who believed Franklin's account. Deborah Dixon, an accomplished reporter with WKRC-TV in Cincinnati, had been following Franklin's trail ever since the murders of teens Darrell Lane and Dante Evans Brown in 1980. She had gone down to Mobile, Alabama, to learn more about Franklin and then to Florida after he was arrested. She later interviewed him twice in prison. During one of those interviews, he told her there was a man in prison in West Virginia for the Rainbow Murders that Franklin said he was responsible for. Her investigative work found its way to the *60 Minutes II* producers at CBS, who aired a story on the connection Dixon had established between Franklin and the Santomero-Durian murders and how Jacob Beard was likely innocent. This led to renewed interest and national attention for the Rainbow killings.

The lead investigator would not abandon his theory of the case, the one the jury had confirmed with its conviction. During a hearing on a motion for a new trial, the lead investigator said he had attempted to get in touch with Franklin back in 1984, although he had no record of his attempt(s). He also admitted that my original fugitive assessment had been made available to him, but he had not read it. After learning this, I checked my notes on the case and found we had consulted on the case in March 1984, received details from the state police, and furnished them with our information. I doubt if any of our material would have changed his mind if he had read my report, but along with DiFalco's report and Dixon's interview, it might have made him realize that the Rainbow murders were very much in Franklin's wheelhouse.

In January 1999, the trial judge, Charles Lobban, reviewed Franklin's several confessions, particularly the one to Melissa Powers, set aside the conviction, and decreed there was reason to grant Beard a new trial. Beard was released on bond from prison and the prosecutor was given until February 11 to decide whether to go forward. The prosecutor and sheriff still believed Beard was the killer and forged ahead with a re-prosecution.

Beard's new trial began on May 16, 2000, in Braxton County. Jurors watched a two-hour video deposition of Franklin. "One of them told me she had dated Blacks . . . and the other one told me she would if she had a chance, so I just decided to waste them at that time," Franklin stated. His veracity was called into question when he said he dumped the bodies "no more than fifteen minutes at most" from Interstate 64, when, in fact, they were found more than an hour away. His video testimony also contradicted witnesses who had said they saw Beard shoot the two women. But several police officers, testifying for the defense, said the witnesses had contradicted themselves a number of times.

The trial lasted a little more than two weeks. On May 31, after less than three hours of deliberation, the jury returned a verdict of not guilty.

Beard's attorney Stephen Farmer brought suit against the West Virginia State Police, the Pocahontas County Police Department, and the prosecutors, claiming his client's civil rights had been violated through the coercion of witnesses and ignoring physical evidence that didn't support their case. Beard was awarded a two-million-dollar settlement. He told reporters no amount of money could make up for his nearly six years in prison.

Those same authorities saw no point in trying Franklin for the murders. He never went to trial for the Rainbow murders, and the chief investigator and sheriff continued to believe in Beard's guilt, as did the prosecutor until he died.

For my part, I strongly believe Franklin was "good" for the murders. They fit both his signature of picking up female hitchhikers and profiling them on their choices, and his M.O. of shooting them in remote areas when they failed his racial purity test. Over the years, Franklin confessed with consistent details to Melissa Powers, Carol Ellis, and Deborah Dixon, among others, and he was more likely to deny crimes he had already admitted to than to admit to crimes he had not perpetrated.

Linkage blindness, as we've said, is an ongoing challenge for investigators, but so is confirmation bias—the attitude and approach with which one goes into a case, embracing evidence that supports that bias and depreciating evidence that does not. And when that happens, someone always pays the price—in this case, Jacob Beard and his and the victims' families and friends.

The Rainbow murders case is a two-decade-long personal tragedy and legal horror story, a searing example of how a misguided investigation and prosecution can ruin lives. But one of the reasons Franklin affected me so deeply is that this was little more than a footnote to his murderous career—just another incidence of picking up hitchhikers and deciding they did not deserve to live. In evaluating crimes, we think about means, motive, and opportunity. Franklin was versatile and adaptable enough that he was able to shift his means and take advantage of diverse opportunities. His motive never changed.

Looking back, the reign of terror that Franklin wrought was clearly far greater than anyone, including myself, could have initially imagined. One of the things we knew about Franklin when we first became aware of him was that he'd been a highly mobile killer. As it turned out, this had proven perhaps his greatest asset. He'd killed over such a large area, during such a broad stretch of time, that many of his crimes and methods had been difficult to link together definitively given his different methods and victimology.

As my retirement from the Bureau loomed, I was ready to finally put Franklin out of my mind.

He had other ideas.

CHAPTER 21

Once out of the Bureau for several years, consulting independently on cases and working on books about profiling and criminal investigative analysis with Mark, I hadn't thought much about Franklin for a while, other than reading that he had forsaken his racist and anti-Semitic views. I hoped it was true. Sitting in a prison cell year after year, the one thing you have is time to think.

He had previously written complimenting Mark and me on our analysis of him in our book *The Anatomy of Motive*. Then, around the beginning of 2001 I received a letter from him, sent to the post office box we kept. It concerned a section in our previous book *Obsession*, about the 1931 Scottsboro Boys case in Alabama, in which nine young African American male teens were arrested for raping two white teenage girls on a freight train, despite the fact that there was no evidence against them other than the word of the two girls, who showed no signs of physical harm and used the charges as a way to stay out of trouble themselves.

In any event, it was this narrative in our book that Franklin took issue with. The letter was neatly written, and the spelling was perfect. Though we've sanitized the N-word, he certainly did not. He wrote:

Dear John,

*Greetings. It's been awhile since I've heard anything from you. An inmate here let me check out a copy of "Obsession" he had, and I was pissed off by the position you're taking regarding the two young white women who were raped by a bunch of n*****s in Scottsboro, AL. One good and easy test for you would be to let your daughter, who I guess is 21 now, get on the empty boxcar of a freight train with about as many n*****s who raped those other girls, and ride with them for about a couple of hundred miles or so, to see if they rape <u>her</u>, and you could let one of your daughter's friends go with her, to make the rest similar to the first case. You could select black students, workers, or whatever, they wouldn't necessarily have to be street n*****s like the Scottsboro Boys, and see what happens. Would you be willing to do that? If not, you're one of the biggest phonies I've ever met, and what you're saying is absolute hogwash!*

Sincerely,

Joseph P. Franklin

*P.S. I'm curious—why do you think two white women would make up a story about getting raped by n*****s, just to avoid getting arrested for vagrancy, like you*

> *stated in your book? It doesn't make sense—how would they have gotten arrested by the police in the first place?*

HE ALWAYS SIGNED HIS NAME WITH A FIVE-POINTED STAR INSTEAD OF A DOT over the "I" in Franklin.

Reading this letter and sharing it with Mark, a harsh reality hit me: I was never going to be free of Joseph Paul Franklin; he had invaded my psyche and intended to reside there. I was instantly reminded of the quote from the German philosopher Friedrich Nietzsche that had become a watchword and warning in my Investigative Support Unit: *Whoever fights monsters should be careful lest he thereby become a monster. And when you gaze long enough into an abyss, the abyss will gaze back into you.*

Though he was safely locked away, I felt Franklin had violated me and my family, that he was still challenging my values and my perception of decency, just as he had attempted to do for the entire nation. He had been in my professional life almost from the beginning of my profiling career, up through my retirement from the Bureau and beyond.

But when I tried to shift my mind back to investigator mode, what I found interesting about the letter—after trying to put aside my feelings of being creeped out over the notion of Joseph Paul Franklin even *thinking* about my daughter—was how he had reacted to the narrative in *Obsession*.

Anyone who has studied the Scottsboro case would come to the same conclusion Mark and I did—that the defendants had been falsely accused, were all innocent and the victims of the racial bigotry of the times. This is established fact, based on solid evidence. But in reading the details, Franklin imposed his own overlay: in any situation in which a white girl or woman accuses a Black boy or man of sexual assault, it must be true, and the African American must be guilty. That was the mythology he had grown up with, it was the narrative that had shaped his life and given it meaning, and he was not about to abandon it at this point. In some ways, it was the same as a sexual serial killer emotionally sustaining himself in prison by reliving the thrill and power of his crimes over and over again in his mind. Clearly, I represented the authority of law enforcement and the FBI to him, the establishment he had fought his whole life. And if I was going to have my say on such a racially charged subject, so was he. I've said repeatedly that you can lock up the body, but you can't lock up the mind.

I didn't hear from Franklin again until April 2004, when he wrote asking me to help find his first wife, Bobbie Louise Dorman. Whether he had reached out to anyone else, I didn't know, but he said he had gotten ahold of an FBI report from 1980 or 1981 that listed her remarried name. I checked the case file and he was correct—that information was on a timeline that had been supplied to me when I worked up the fugitive assessment in 1980. I have no idea how Franklin got a copy.

It wasn't infrequent that incarcerated killers reached out to me by mail. Part of it was probably their knowledge of all the prison interviews I'd conducted, which showed that I was willing to listen to them when few law enforcement officials were. Part of it was the books Mark and I had published, which had reached a wide audience, apparently even in prison.

Franklin said he wanted to get back in touch with Bobbie, though he added, "John, please make it clear to her that I'm not trying to get a serious relationship going again. I just want to kind of, like, keep in touch, OK?"

What was pathetic about this was that as his life presumably neared its end, Franklin must have realized there was nothing left for him, no human connection. If I was the best person he could think of to help reach his ex-wife, he must truly be desperate. Even with the most hardened criminals, I've found there is generally some sentimental core. With Franklin, it was the daughter to whom he'd never been a father and the wife to whom he'd only briefly been a husband. But they took a back seat to the pursuit of his murderous hatred.

It's hard to imagine an emptier life.

AS IS SO OFTEN TRUE IN DEATH PENALTY CASES, THE APPEALS PROCESS WAS lengthy. While his case wound its way through various levels of state and federal appeals, Franklin began to express remorse for the first time. He said he had been mentally ill at the time of the Gordon shooting and told anyone who would listen that he had forsaken his anti-Black and anti-Jewish views and come to believe in the equality of all of God's children. I hoped he was sincere, though my encounters with him over the years made me retain a healthy dose of skepticism.

Finally, on August 14, 2013, the court set the date for the now sixty-three-year-old Franklin to be executed by lethal injection on Wednesday, November 20. He had been thirty when he was finally captured in 1980; he had spent more than half his life in prison.

The determination of Franklin's execution date coincided with a decision by the Missouri Department of Corrections to switch to a one-drug execution protocol employing pentobarbital, a short-acting barbiturate sedative that used to be marketed as a sleeping pill under the trade name Nembutal. Up until then the state had employed a standard three-drug series of sodium thiopental, pancuronium bromide, and potassium chloride, all of which were becoming increasingly difficult to obtain. Missouri then decided to switch to high doses of propofol, commonly used in hospitals for anesthesia. But the European Union, which opposed the death penalty, threatened to withhold supplies to the United States if it was used in execution protocols. In high doses, pentobarbital shuts down the heart and respiratory system, leading to death.

As the execution date neared, Larry Flynt came out against it, saying that he didn't believe in the death penalty and felt a lifelong prison sentence would be greater and more effective punishment, "far harsher than the quick release of a lethal injection."

Franklin reached out by phone to a number of the reporters he had spoken to over the years, calling himself a changed man who no longer hated African Americans or Jews and had repented to God. He had gotten to know Blacks as individuals in prison, and he was no longer anti-Semitic, but pro-Semitic. "It just goes to show you how you can be totally convinced that something you believe is true when nothing can be further from the truth."

Back in February 1997, while he awaited formal sentencing in the St. Louis County Jail for the Gerald Gordon murder, the one that would finally bring him the death sentence he said he wanted, Franklin shared with Kim Bell of the *Post-Dispatch* his contempt for convicted killers who try to avoid execution: "It disgusts me that these guys try to save their miserable lives."

Now, his attorney Jennifer Herndon said, "He believes he should be kept alive so he could help other people overcome their racist views."

There is some karmic justice in his own last-minute efforts to stave off the lethal injection. In truth, though, Franklin never actually claimed he wanted to die. He simply preferred to die later in a state prison than sooner, as he thought he would, at the hands of Marion's guards or Black inmates. What he actually preferred was to be freed from jail, either by escape or when the race war was finally won and he was greeted as a hero of the revolution.

He was moved from Potosi the short distance to the Eastern Reception, Diagnostic and Correctional Center in Bonne Terre, where executions took place. His attorneys and the American Civil Liberties Union attacked the protocol, claiming not enough was known of the effects of pentobarbital to ensure against "an excruciatingly painful execution," as Franklin's attorneys put it, potentially violating the Eighth Amendment ban on cruel and unusual punishments.

"There are simply too many unanswered questions to justify ending someone's life," ACLU of Missouri's executive director, Jeffrey A. Mittman, wrote in an email on Friday, November 15.

Officials pointed out that pentobarbital was commonly used to euthanize ailing pets. Franklin took issue with that, too, and not for the reasons his lawyers or the ACLU did. He told *Post-Dispatch* reporter Jeremy Kohler, "It's humiliating to put a person to death with a drug like that. It's humiliating to put someone like me in the same category as an animal. It isn't moral to kill somebody using that type of drug. I don't think it is right." I think the irony speaks for itself.

On Monday, November 18, Governor Jay Nixon denied clemency. He asked Missourians to keep Gerald Gordon, his family, and all of Franklin's other victims and their survivors in their thoughts and prayers.

Lee Lankford, retired chief of the Richmond Heights PD, who had seen the case through from beginning to end, commented, "How many lives did he take on his rampage across the United States? Just to be put to sleep, that's the easiest way out of here." Having looked at so many crime scene photos and read so many medical examiners' reports, I've often felt the same way myself.

Franklin refused a final meal, asking that it be given to a hungry child or homeless person instead. On the morning of November 20, Franklin was brought from his holding cell into the execution chamber and strapped to a table. He offered no resistance and said nothing.

At 6:05 A.M., Governor Nixon gave the okay for the execution to commence.

Franklin made no final statement. At 6:07 he was injected with five grams of pentobarbital in a 5 percent solution. Once the lethal chemical began running through his veins, he was observed to swallow hard, breathe heavily for several moments, then lie still. Three witnesses from the media reported that he did not appear to be in pain. The entire process took about ten minutes.

Perhaps it was Hal Harlowe, the former Dane County district attorney who prosecuted the Manning-Schwenn murders in Madison, Wisconsin, who gave Franklin his most truthful epitaph. Upon learning in 1997 that the Missouri jurors had handed down a death sentence, Harlowe noted: "He was ordinary and not very bright. He was not nearly as special as the many people he killed."

EPILOGUE

"Where will he go next, this phantom from another time, this resurrected ghost of a previous nightmare—Chicago; Los Angeles; Miami, Florida; Vincennes, Indiana; Syracuse, New York? Anyplace, everyplace, where there's hate, where there's prejudice, where there's bigotry. He's alive. He's alive so long as these evils exist. Remember that when he comes to your town. Remember it when you hear his voice speaking out through others. Remember it when you hear a name called, a minority attacked, any blind, unreasoning assault on a people or any human being. He's alive because through these things we keep him alive."

—Rod Serling, closing monologue from "He's Alive,"
The Twilight Zone

When I learned that Joseph Paul Franklin had been executed, I felt that justice had finally been rendered in his case, though I can't say I rested much easier. He was dead, but the legacy of hatred, intolerance, and resentment he hoped to encourage was still alive, as it is to this day.

And it is why the Franklin case continued to haunt me, and why understanding a killer like Franklin is important—and urgent. Hate always has an antecedent and a target—it comes from somewhere and it goes somewhere. With all of the serial killers I have studied we confront the ongoing question: Are they born or made? Is it nature or nurture? The answer, as we have seen, is *both*; in fact, a dynamic interaction between the two. Despite his grisly success as a repeat killer, Franklin is not unique and the shadow he cast is long.

In 1978, in the midst of Franklin's murderous reign of terror, a rabidly racist and anti-Semitic screed of a novel titled *The Turner Diaries* was published. Set in the near future, the book centers around one Earl Turner, a white revolutionary who joins "the Organization," an Aryan supremacy group, to wage guerilla war against the repressive American government, referred to as "the System." As part of its campaign, the Organization stages the "Day of the Rope" in Los Angeles, during which "race traitors" are publicly hanged. Turner dies a hero when he flies a small plane equipped with a nuclear warhead into the Pentagon. An epilogue tells how, over the next century, the Organization has triumphed and eliminated all of the nonwhite races, as well as the Jews.

The novel went on to sell more than half a million copies and, among other acts of domestic terror, is known to have inspired a shady white supremacist group known as the Order in the 1984 murder of liberal Denver talk-show host Alan Berg and the 1995 Oklahoma City federal building bombing by Timothy McVeigh, which mirrored the novel's bombing of FBI headquarters. Pages from the novel were found in McVeigh's getaway car.

The *Turner Diaries* author is listed as Andrew Macdonald,

actually a pseudonym for William Luther Pierce III, a physicist by profession who had taught at Oregon State University. Pierce's life's work, however, was as a professional purveyor of hate. A onetime member of the John Birch Society and National Socialist White People's Party, he founded the white supremacist National Alliance in 1974. As John Sutherland wrote in the *London Review of Books*, "*The Turner Diaries* is not the work of a Holocaust-denier (although Pierce gives us plenty of that) so much as a would-be Holocaust-repeater."

Pierce's next novel, *Hunter*, also written under the Andrew Macdonald pseudonym, came out in 1989. It follows Oscar Yeager, who sets out on a campaign to assassinate interracial couples and civil rights advocates and settle "the Jewish question." There is little doubt on whom *Hunter* is based, and Pierce dedicated it to Franklin, "the Lone Hunter, who saw his duty as a White man and did what a responsible son of his race must do, to the best of his ability and without regard for the personal consequences." Like *The Turner Diaries*, *Hunter* is considered an urtext of the movement.

Due to the remarkable advances in communications technologies, we find ourselves living at a time when it is far easier to radicalize and inspire hate than ever before. The internet and social media have made it much easier to spread philosophies like Franklin's than it was in his time, and he would undoubtedly be delighted to see his face appearing frequently on sites like YouTube. With today's technology, conversations that used to be limited to basements and meeting halls, the kinds of words and places that first helped radicalize Franklin, now have tens of thousands of participants online. Corrosive ideas, hate speech, conspiracies, and even potential crimes have a home online unlike any they have ever known. As early as 2000, the Southern Poverty Law Center had identified as many as five hundred racial hate websites, a number that has exploded in the decades since, metastasizing across social media platforms as well.

With the proliferation and reach of these online spaces today, I am reminded of how Franklin told me he used white supremacist newspapers the way many sexual predators use violent pornography, to fuel both his fantasy of racial violence and his belief that he had a heroic role to play within a larger movement. The internet today is rife with spaces that do precisely the same thing, incubating and disseminating hate-fueled lies and conspiracies.

And there clearly is a point when dangerous speech turns into actual danger; when a switch flips and suddenly talk is no longer enough. Just as Franklin tired of the talk of the hate groups he was a part of, so do similar white supremacists of today.

On the evening of June 15, 2015, a slightly built, shaggy-haired twenty-one-year-old unemployed high school dropout named Dylann Storm Roof walked into the historic Emanuel African Methodist Episcopal Church at 110 Calhoun Street in Charleston, South Carolina, long associated with the civil rights movement. He was wearing a gray sweatshirt and jeans and asked one of the parishioners for the pastor, Clementa C. Pinckney. When told Reverend Pinckney was attending a Bible study meeting, Roof went in and sat down next to him. He participated for a while and later said the other participants were very nice to him.

At approximately 9:05 P.M., Roof stood up, took a Glock 41

semiautomatic .45-caliber handgun from his fanny pack and opened fire around the room. He killed nine people, six women and three men, all African Americans aged twenty-six to eighty-seven, including forty-one-year-old Pinckney, a Democratic member of the state senate. His wife and two daughters were in the church building at the time.

Roof ran from the scene and was captured the next morning at a traffic stop in his hometown of Shelby, North Carolina, 245 miles away. He waived extradition and was returned to South Carolina. At a bond hearing he attended from jail by videoconference, survivors and relatives of five of the victims told him they forgave him and were praying for his soul. In September, he agreed to plead guilty to multiple state murder charges in exchange for a life without parole sentence instead of the death penalty urged by Governor Nikki Haley.

On December 15, 2016, at the end of a federal trial in which one of his attorneys would not allow a guilty plea because of a possible death penalty, Roof was found guilty of all thirty-three murder and hate crimes charges against him. He was sentenced to death on January 10, 2017. By that point, like Franklin, he had dismissed his attorneys and chose to represent himself.

In January 2020, a new defense team filed a 321-page motion with the United States Court of Appeals for the Fourth Circuit, seeking to overturn the federal conviction and death sentence on the grounds that Roof should not have been allowed to represent himself because he was "disconnected from reality" and suffering from "schizophrenia-spectrum disorder, autism, anxiety and depression."

This was in opposition to a statement he made at the hearing, contradicting his attorneys, as Franklin often did, and asking the court to disregard them: "There's nothing wrong with me psychologically."

In a jailhouse journal he wrote, after the attack, "I am not sorry. I have not shed a tear for the innocent people I killed."

What some people find so difficult to conceive is that someone may make a statement like this not because he is crazy but because this is what he believes. Is Roof mentally ill? I would say so, just as I believed about Franklin. But did he know right from wrong and was he able to control his actions? Just like Franklin, absolutely.

And here is the gist of it: His lawyers said that Roof was unconcerned with both the life without parole and death sentences because he had killed with the idea of starting a race war. And after victory in that war, he would be freed by the victorious white supremacists. In his time, Franklin believed the same.

"I did what I thought would make the biggest wave," Roof wrote from prison, "and now the fate of our race is in the hands of my brothers who continue to live freely."

As stated in an in-depth December 2019 article on hate crimes in *The Washington Post*, "Roof has become a cult figure among white supremacists, especially those who espouse racial violence." Just as Franklin was before him.

Both Franklin and Roof were delusional in the sense of expecting a race war, after which they would be embraced as heroes. But both these useless losers were sane and rational, within their own belief systems, when they took lives of people they considered inferior.

On March 20, 2017, a jobless twenty-eight-year-old army veteran named James Harris Jackson, known by his middle name, fatally stabbed sixty-six-year-old Timothy Caughman shortly after 11:00 p.m. on Ninth Avenue in Manhattan. Caught on surveillance footage, Jackson told NYPD detectives the next day that he had been inspired by Dylann Roof's act and that the stabbing was "practice" for a larger racial slaughter in Times Square. The night before the attack he had typed a manifesto he entitled "Declaration of Total War on Negros."

Unlike other right-wing extremists, Jackson came from a liberal and progressive Baltimore family that supported Barack Obama for president, and he attended a Quaker Friends school. His grandfather had been instrumental in desegregating the Shreveport, Louisiana, school system. Perhaps Jackson's evolving racial hatred was a means of rebellion.

I see Dylann Roof as Franklin's spiritual son—a young man whose hate and resentment at his own fate was so strong that he not only had to blame others, he had to translate his hate into action. And I see Harris Jackson as Roof's. As I said regarding Franklin, perhaps intense intervention along the way could have turned either of them around. But it's too late for that now.

Joseph Paul Franklin has spawned many spiritual children.

On August 11 and 12, 2017, white supremacists marched through the college town of Charlottesville, Virginia, in a demonstration they called the Unite the Right rally. Carrying lit torches, they chanted "Jews will not replace us!" and "Blood and Soil," a slogan that originated in Nazi Germany. Some of the marchers waved Nazi flags while others wore red Donald Trump "Make America Great Again" caps. The ostensible purpose of the rally was to protest the dismantling of a statue of Confederate general Robert E. Lee from a city park.

On the second day, James Alex Fields Jr., a twenty-year-old self-identified white supremacist, intentionally rammed his car into a crowd of counterprotesters, killing thirty-two-year-old Heather Heyer, a paralegal, and injuring nineteen others.

David Duke, former grand wizard of the Ku Klux Klan, called the rally a "turning point." Vox.com quoted him as saying, "We are going to fulfill the promises of Donald Trump. That's what we believed in. That's why we voted for Donald Trump, because he said he's going to take our country back."

For his part, President Trump, who had frequently been accused of fomenting divisiveness and encouraging hate speech, commented on the protesters, "You had some very bad people in that group, but you also had some people that were very fine people, on both sides."

At about 9:50 on Saturday morning, October 27, 2018, forty-six-year-old Robert Gregory Bowers walked into the Tree of Life Congregation in the peaceful Squirrel Hill neighborhood of Pittsburgh during a sabbath service attended by about seventy-five Jewish worshippers. The bearded, heavyset white male opened fire with a Colt AR-15 semiautomatic rifle and three Glock .357 semiautomatic pistols. During a twenty-minute barrage he killed eleven people, aged fifty-four to ninety-seven, and wounded six others, including four Pittsburgh police officers.

According to *USA Today*, he shouted "All Jews must die!" as he fired at random.

He surrendered to police after being wounded by the SWAT team, two of whose members he first wounded in a shootout as they entered the building.

A white supremacist and neo-Nazi, Bowers believed Jews to be "the children of Satan." He was particularly upset at HIAS, the 130-plus-year-old Hebrew Immigrant Aid Society, for helping "to bring invaders in that kill our people."

On Friday, November 1, 2019, the FBI arrested a man in Colorado, and charged him with plotting to blow up Temple Emanuel synagogue in Pueblo. In a private Facebook message intercepted by undercover agents, he wrote "I wish the Holocaust really did happen. The Jews need to die." Some of the congregation's members were children of Holocaust survivors.

Sadly, the list seems to grow with each passing month, and by the time you are reading these pages, there probably will have been even more. What all of these individuals share with Joseph Paul Franklin is that they are "lone wolves," men who embrace the collective philosophy of the hate group but take it one shattering step further and act on it. Men for whom words are no longer enough. Those who carried the torches and spouted the racist slogans in Charlottesville may talk the talk and even walk the walk, but one of them, James Fields, took it upon himself to drive his car into a peaceful gathering.

Thoughts and words matter. They have power—for both good and evil. They inspire some to violence, and those, in turn, inspire others. The good news is that today, the same online visibility that enables hate messages to flourish also makes it more possible for us to spot those who may be teetering on the edge and intervene before they move to action. And it is encouraging that in the midst of the COVID-19 pandemic, people of all races, faiths, and ethnicities were willing to come together to demonstrate in large numbers, all across the country, supporting the civil and human rights that those like Franklin seek to deny.

The journey to reckon with our nation's searing history of racial hatred, intolerance, and discrimination is ongoing, and there are no neutrals in that struggle. The shadow cast by Joseph Paul Franklin and his like is long and dark, so the sunlight to eradicate it must be even brighter and stronger.

ACKNOWLEDGMENTS

Once again, our special and heartfelt thanks go out to:

Our wonderful and discerning editor, Matt Harper, whose talent, insight, and perspective guided us every step of the way; and the entire HarperCollins/William Morrow/Dey Street family, including Anna Montague, Andrea Molitor, Danielle Bartlett, Bianca Flores, Kell Wilson, and Beth Silfin.

Our amazing researcher Ann Hennigan, who has worked with us since the beginning and is an integral part of the team. She was critical in helping us keep this complicated story straight and her perceptions added mightily.

Our ever supportive and resourceful agent, Frank Weimann and his team at Folio Literary Management.

John's colleagues at Quantico, with special remembrance of the late Special Agent Roy Hazelwood and late Secret Service Special Agent Ken Baker, two of the very best.

Mark's wife, Carolyn, among many other attributes, our Mindhunters chief of staff and in-house counselor.

We are also indebted to our British friend Mel Ayton, whose book *Dark Soul of the South: The Life and Crimes of Racist Killer Joseph Paul Franklin* proved an invaluable resource. So, too, was the voluminous journalistic record over a forty-year period, with special appreciation for the Associated Press, United Press International, and Mark's old newspaper, the *St. Louis Post-Dispatch*.

Finally, though this book is dedicated to all of Franklin's known murder victims, we also want to pay respectful tribute to their noble and courageous survivors, family, and friends. A murderer's bullet, knife edge, or bomb always strikes a multitude of targets and the ripples of that disruption move out in concentric circles for generations. They will always have a firm place deep in our hearts.

ABOUT THE AUTHORS

JOHN DOUGLAS is a former FBI special agent, the bureau's criminal profiling pioneer, founding chief of the Investigative Support Unit at the FBI Academy in Quantico, Virginia, and one of the creators of the *Crime Classification Manual.* He has hunted some of the most notorious and sadistic criminals of our time, including the Trailside Killer in San Francisco, the Atlanta Child Murderer, the Tylenol Poisoner, the Unabomber, the man who hunted prostitutes for sport in the woods of Alaska, and Seattle's Green River killer, the case that nearly ended his own life. He holds a doctor of education degree, based on comparing methods of classifying violent crimes for law enforcement personnel. Today, he is a widely sought-after speaker and expert on criminal investigative analysis, having consulted on the JonBenet Ramsey murder, the civil case against O. J. Simpson, and the exoneration efforts for the West Memphis Three and Amanda Knox and Raffaele Sollecito. Douglas is the author, with Mark Olshaker, of seven previous books, including *Mindhunter*, the Number 1 *New York Times* bestseller that is the basis for the hit Netflix series.

MARK OLSHAKER is a novelist, nonfiction author, and Emmy Award–winning filmmaker who has worked with John Douglas for many years, beginning with the PBS *Nova* Emmy-nominated documentary *Mind of a Serial Killer.* He has written and produced documentaries across a wide range of subjects, including for the Peabody Award-winning PBS series *Building Big* and *Avoiding Armageddon.* Olshaker is the author of highly praised suspense novels such as *Einstein's Brain*, *Unnatural Causes*, and *The Edge*. In the other realm of life-threatening mysteries, he is coauthor with Dr. C.J. Peters of *Virus Hunter: Thirty Years of Battling Hot Viruses Around the World*, and with Dr. Michael Osterholm of *Deadliest Enemy: Our War Against Killer Germs.* His writing has appeared in *The New York Times*, *The Washington Post, The Wall Street Journal*, *USA Today*, the *St. Louis Post-Dispatch*, *Newsday*, *Time*, *Fortune*, and *Foreign Affairs.*

Both authors and their wives live in the Washington, D.C., area.